Fabrication Characteristic Techniques of Heat-Treated Metals

Sheerin Bano

TABLE OF CONTENTS

Chapter No.		Contents	Page No.
1	**Introduction**		**1**
	1.1	Overview	1
	1.1.1	Wrought Alloys	3
	1.2	Al7068	6
	1.3	Mechanical Properties of Al7068	8
	1.4	Comparison of Mechanical Properties of Al7075 vs Al7068	10
	1.5	Reinforced Nanocomposites	11
	1.6	Effect of Micro and Nano Reinforced on composites	13
	1.7	Reinforcement of micro particles with Silicon Carbide	17
	1.8	Reinforcement of micro particles with Silicon Nitride (Si_3N_4)	18
	1.9	Micro SiC (Silicon Carbide) Particles	20
	1.10	Micro Si_3N_4 (Silicon Nitride) Particles	22
	1.11	Organization	25
	1.12	Chapter Summary	26
2	**Literature Review**		**27**
	2.1	Introduction	27
	2.2	Literature Review on Mechanical Properties & Fabrication methods	27
	2.3	Literature Review on Al7068	58
	2.4	Outcomes of Literature Review	66

	2.5	Motivation	67
	2.6	Strength of Study	68
	2.7	Objectives of Research	69
3	**Methodology and Characterization**		**70**
	3.1	Materials and Processing Route	70
	3.2	Fabrication and Characterization Techniques	70
	3.3	Methodology	76
	3.4	Preparation of Standard (ASTM) Specimen	84
	3.5	Proposed Approach for Alloy Synthesis	86
	3.6	T6 Heat Treatment	87
4	**Results and Discussion**		**89**
	4.1	Investigation of Mechanical Properties	89
	4.1.1	Investigation of Tensile Test	90
	4.1.2	Charpy Impact Testing before shot Peening	94
	4.1.3	Charpy Impact Testing after shot Peening	97
	4.1.4	Rockwell Hardness Testing	100
	4.2	Material Characterization	103
	4.2.1	Field Emission Scanning Electron Microscopy	103
	4.2.2	Optical Microscopic Analysis	111
	4.2.3	Fractographic Analysis of Tensile Fractured Surface	116
5	**Conclusion and Future Scope**		**119**
	5.1	Conclusion	119
	5.2	Future Scope	124
	5.3	Social Impact	125
	5.3.1	Advancement in Material Science:	125
	5.3.2	Enhanced Mechanical Properties:	125
	5.3.3	Potential for Light Weight:	125

Chapter - 1
Introduction

Chapter 1
Introduction

1.1 Overview

Engineering materials are being played important role as well as shaping human civilization with continuous development and our focused on improving efficiency, performance, and the responsible use of resources. In today's world, the shortage of natural resources has been increased now these days, so more important innovation in material science and engineering is required. At earlier times, industrial growth mainly depended on the inventions of new materials as well as the improvement of existing material, which brought major changes across different industries. For instance, the automobile sector has evolved significantly, shifting from heavy and inefficient vehicles to lighter, more fuel-efficient, eco-friendly, and advanced designs, showing the importance of material development.

The iron, copper as well as its alloys have been used since thousands of years, aluminum is still challenge for engineers because it is one of the most abundant elements found in the earth's crust, firstly found in the form of bauxite ore. Mostly people use aluminum in the last century as well as its properties modification make as a good alternative compared to other steel due to unique mechanical properties, such as low density, high specific strength, excellent corrosion resistance, and ductility. Therefore, aluminum alloys have been using in wide applications of aerospace, defense, transportation, agriculture, and chemical industries, and its superior mechanical performance characteristics directly contribute to technological advancement and industrial growth. [1]

Aluminum has become the most broadly using non-ferrous metal worldwide, even better and cheaper than copper, due to its unique properties of low density, excellent corrosion resistance, and high recyclability. Transportation sector is the notable example of this field. Earlier generations of automobile and aerospace is demanded greater fuel consumption and required large maintenance due to by heavier weight and less efficient material design. With the advances of engineering material and manufacturing technologies, the automotive industry has shifted towards the lightweight and energy-efficient solutions. This shift contributed more fuel savings, lowers maintenance costs, and supports the wider goal of environmental sustainability. The 21st century marked a key move to all-aluminum alloy automotive designs, a landmark in engineering material innovation. [2]

In order to overcome aluminum's inherent mechanical deficiencies, mixing it with other materials has been vital to its development. By combining specific alloying elements such as zinc (Zn), copper (Cu), silicon (Si), manganese (Mn), and magnesium (Mg), the mechanical strength, stiffness, and overall performance of aluminum can be considerably improved. These alloying compositions additions enable the aluminum alloys to achieve higher strength-to-weight ratios, improved wear resistance, and greater adaptability across diverse engineering applications. Consequently, aluminum alloys now serve as very important materials in aerospace, defense, automotive, and industrial sectors, where its tailored mechanical and physical properties offer a stability between performance, durability, and sustainability.[2]

Among the wide range of aluminum alloys, Al-Si alloys are extensively utilized in IC engines due to its desirable combination of properties, such as low coefficient of thermal expansion, enhanced bearing characteristics, and corrosion resistance, while maintaining sufficient strength. Al-Zn-Mg alloys, are having use in structural components for both automotive as well as for aerospace sectors due to its good mechanical strength. Similarly, Al-Mg casting alloys shows outstanding machinability, enhanced corrosion resistance, as well as an anodized finish, that significantly expand its use in engineering applications.

Aluminum alloys are classified into two types such as wrought alloys and cast alloys, depending on its manufacturing methods. Wrought alloys are prepared into semi-finished products such as sheets, foils, and extrusions, while cast alloys are mainly manufactured in the as-cast condition. Alloys are further divided into heat-treatable and non-heat-treatable conditions under the wrought category. Non-heat-treatable alloys have been achieved strength through the mechanisms of solid-solution hardening and strain hardening after annealing. This strengthening mechanism depends on alloying elements like copper (Cu), manganese (Mn), silicon (Si), and nickel (Ni), which have high solubility in aluminum at elevated temperatures but limited solubility at room temperature.[2]

1.1.1 Wrought Alloys

The wrought heat-treatable aluminum alloy consists of major series like 2XXX series (Al-Cu, Al-Cu-Mg compositions), the 6XXX series (Al-Mg-Si compositions) as well as 7XXX series (Al-Zn-Mg and Al-Zn-Mg-Cu compositions). This alloy systems are the backbone of modern lightweight engineering materials, mainly in the aerospace and automotive industries, where the dual demands of high strength-to-weight ratio and resistance against environmental degradation is essential.[3]

Pure aluminum (1XXX series) invariably contains small amounts of iron and silicon. Due to the very limited amount of Fe in Al, secondary intermetallic compounds such as $FeAl_3$, Fe_3SiAl_{12}, and $Fe_2Si_2Al_9$ are commonly observed within the microstructure of as-cast alloys, also with metastable phases (e.g., $FeAl_6$) formed during rapid solidification. The presence of minor amount element like Cu or Mn, in pure aluminum series may alter the stability, morphology, and distribution of these phases. Advanced characterization techniques are required to identify these phases, while subsequent heat treatments usually convert metastable constituents into equilibrium parts.

In the 1100 aluminum alloy, iron-rich intermetallic compounds constitute the primary microstructural constituents, and its distribution is significantly changes by subsequent heat treatments process and mechanical working as shown in table 1.1. The annealed sheet shows a nearly equiaxed grain structure, while the volume fraction of second-phase particles is largely indicated by iron content.[3]

Aluminum–manganese alloys (3XXX series) are mainly derived from 1100 aluminum by the additions of Mn. In aluminum–manganese alloys, the as-cast microstructure is primarily named by $(Mn,Fe)Al_6$, which transforms into $(Fe,Mn)_3SiAl_{12}$ by the heating. Manganese also supersaturates the solid solution and after precipitates as fine dispersoids. These dispersoids shows elongated or flattened grains after annealing process, and partial Mn solubility remains; With the addition of magnesium, the phase balance is further shifted toward $(Mn,Fe)Al_6$, due to its affinity for Si.

The 5XXX series aluminum–magnesium alloys mainly keep magnesium in solid solution in wrought products, while increasing the magnesium concentration in as-cast ingots causes the production of eutectic Mg_2Al_3 and Mg_2Si phases, during low-temperature treatments, Mg_2Al_3 or its metastable form may precipitate within grains boundaries when magnesium levels exceed above approximately 3.5% as shown in table 1.1. Additions of chromium and manganese produce fine dispersoids such as $Cr_2Mg_3Al_{18}$ and $MnAl_6$, which refine the microstructure. Cold working introduces deformation bands adorned with magnesium-rich precipitates, thereby enhancing strength and corrosion resistance, making these alloys suitable for marine, automotive, and structural applications.[3]

The 4XXX series aluminum–silicon alloys are primarily uses in welding and brazing filler materials, except architectural applications and forged pistons. The performance of this 4XXX series in joining processes depends on maintaining a fine and uniform wrought structure. In the as-cast condition, these alloys mainly contain elemental silicon and $Fe_2Si_2Al_9$ phases, while thermal treatment causes silicon coalesce and spheroidize. For example, a 4343 clad layer on a 3003 core used for brazing.[3]

The 6XXX series aluminum–magnesium–silicon alloys are designed by the solubility of Mg_2Si and thereby utilize precipitation hardening, making highly versatile for structural applications. Depending on alloying elements, the iron-rich phases may be Fe_3SiAl_{12}, $Fe_2Si_2Al_9$, or stabilized $(Fe,Mn,Cr)_3SiAl_{12}$ in the absence of manganese or chromium. For heating cast structure dilute alloys grades such as 6063, at moderate temperature dissolves all Mg_2Si, while in higher-strength alloys like 6061, has an excess of Mg_2Si and precipitates in a Widmanstätten form during slow cooling. Solution-treated 6061 exhibits microstructures that are difficult to differentiate between T4 and T6 tempers without specialized etching methods. Additionally, some of the 6XXX alloys optimized for electrical conductivity are put into an over aged condition.[3]

The 7XXX (Al-Zn-Mg-Cu) alloy derives from the higher solubility of zinc in aluminum itself, but requiring the minor addition of Mg and Cu to form strengthening precipitates such as $Mg(Zn,Cu,Al)_2$ and Al_2CuMg. In cast ingot form, alloy 7075 produces certain compounds, such as $(Fe,Cr)_3SiAl_{12}$ and Mg_2Si. It also forms a special mixture, called a pseudobinary eutectic, made up of aluminum and $MgZn_2$. After addition of copper in place of zinc can be written $Mg(Zn,Cu,Al)_2$. Iron-rich phases transform to Al_7Cu_2Fe. Mg_2Si is relatively insoluble and tends to spheroidize by subsequent heating. After heat treatment (solutionized) wrought alloy contains three compunds Al_7Cu_2Fe, $(Fe,Cr)_3SiAl_{12}$, and Mg_2Si, along with the dispersoid. Due to dispersoid banding recrystallized grains are elongated. The uncrystallized regions consist of very fine sub grains, whose boundaries are lined with hardening precipitates [3]. Table 1.1 show the comparison of aluminum alloy series.

Table 1.1 Comparison of Aluminum Alloy Series

Series	Main Alloying Elements	Key Microstructural Phases	Heat Treatment Response	Typical Properties	Applications
1XXX (Pure Al)	pure Al with small Fe, Si	$FeAl_3$, Fe_3SiAl_{12}, $Fe_2Si_2Al_9$; metastable $FeAl_6$	Limited response; metastable phases convert to equilibrium on heating	Excellent corrosion resistance, high ductility, low strength	Chemical equipment, roofing, siding, packaging
4XXX (Al–Si)	Silicon (primary), minor Fe	Elemental Si, $Fe_2Si_2Al_9$; Si spheroidizes on heating	Limited strengthening; relies on Si morphology control	Good wear resistance, low thermal expansion, moderate strength	Automotive engines, pistons, welding/brazing filler
5XXX (Al–Mg)	Magnesium, with Cr/Mn additions	Mg_2Al_3, Mg_2Si; dispersoids $Cr_2Mg_3Al_{18}$, $MnAl_6$	Non-heat-treatable; strengthening via solid solution + work hardening	High strength-to-weight ratio, excellent corrosion resistance (esp. marine)	Shipbuilding, automotive, pressure vessels, structural
6XXX (Al–Mg–Si)	Mg + Si (balanced for Mg_2Si)	Mg_2Si (precipitates), Fe_3SiAl_{12}, $Fe_2Si_2Al_9$, $(Fe,Mn,Cr)_3SiAl_{12}$	Heat-treatable; precipitation hardening (T4, T6 tempers)	Good strength, corrosion resistance, machinability, weldability	Architectural, automotive, aerospace, electrical conductivity alloys
7XXX (Al–Zn–Mg–Cu)	Zn + Mg + Cu (minor Cr, Mn, Zr)	$Mg(Zn,Cu,Al)_2$, Al_2CuMg, Al_7Cu_2Fe, $(Fe,Cr)_3SiAl_{12}$, Mg_2Si	Heat-treatable; strong precipitation hardening	Very high strength, lower corrosion resistance (needs protection)	Aerospace, defense, high-performance automotive

1.2 Al7068

Aluminium metal matrix composites have emerged as advanced engineering materials due to its high strength, lightweight nature properties compared to other conventional alloys. The mechanical properties of metal matrix composite are strongly influenced by the different types of reinforcement, manufacturing method, and processing parameters. The clear understanding of synthesis routes, reinforcement selection, and resulting mechanical behavior is essential for industry to optimizing for the industrial use. While significant progress has been made in manufacturing techniques and applications, challenges related to large-scale production and commercialization still remain. Addressing these issues will play a key role in expanding the practical adoption of AMCs across diverse industrial sectors.[4]

Aluminum Alloy Al7068 is one of the highest-strength aluminum alloys available, offering remarkable mechanical performance with maintaining a good balance of weight and corrosion resistance. It is extensively used in aerospace, defense, and other high-performance engineering applications because of its superior strength-to-weight ratio. The performance of AMCs mainly depends on the type of reinforcement, the fabrication technique, and the processing conditions. The clear understanding of synthesis routes, reinforcement selection, and resulting mechanical behavior is necessary to improve their performance in industrial applications. Researcher has done significant progress in manufacturing techniques and applications; challenges related to large-scale production and commercialization still remain to solve. Solving these challenges is important for wider use of AMCs in various industries.

Composition: Al7068 mainly consists of aluminum (Al) as the base element, with key alloying elements such as zinc (Zn), magnesium (Mg), and copper (Cu), along with minor additions of zirconium (Zr) and chromium (Cr). This alloy's compositions make it material too stronger to achieve greater tensile strengths exceeding many steels, making it one of the strongest commercially available aluminum alloys.

Properties

- **Ultra-High Strength:** Al7068 is one of the strongest commercially available aluminum alloys, with ultimate tensile strength up to 700 MPa and even greater than of Al7075. Its makes highly competitive grade compared with other grades of steel while remaining much lighter.

- **Lightweight:** Al7068 has low density properties as well as providing an outstanding strength-to-weight ratio, which is essential in aerospace and defense applications.
- **Corrosion Resistance:** The alloy evinces good corrosion resistance at normal environments, though surface treatments as well as coatings having improved the durability in aggressive conditions.
- **Heat Treatable:** Al7068 having good response with heat treatment process, particularly in tempering process such as T6511 as well as improved the strength and toughness according to engineering requirements.
- **Machinability:** Al7068 also having relatively excellent machinability properties compared to other high-strength aluminium alloys as well as it can be used in mechanical precision components with the help of machining tools and techniques.
- **Fatigue Resistance:** Al7068 having its fine grain structure and high strength as well as Al7068 exhibits excellent fatigue resistance, making it suitable for cyclic load-bearing applications.

Applications

- **Aerospace:** It is mainly used in aerospace structures such as fuselage frames, wing fittings, as well as landing gear components, where high strength and low weight are important.
- **Defense & Military:** Used in military hardware such as firearms (e.g., rifle receivers), armored vehicle components, and air defense systems, benefiting by the combination of its alloys strength and reliability.
- **Automotive:** It is used in high-performance automotive components such as suspension systems, drive shafts, and lightweight racing parts, where high strength is needed as well as minimizes over weight.
- **Sports Equipment:** Used in advanced sporting items, like high-strength bicycle parts, archery equipment as well as precision sports gear, its performance depends on alloys strength and lightweight design.

- **Industrial & Engineering:** It is used in high-strength fasteners, pressure vessels, robotics, as well as mechanical parts that are demanding both durability and low weight.
- **Space Applications:** Increasing demand in satellite as well as space vehicle structures where high tensile strength and weight savings directly improve mission efficiency.

1.3 Mechanical Properties of Al7068

The mechanical properties of Al 7068 is one of the strongest commercially aluminum alloys, even better than the performance of Al7075. These properties are highly required in aerospace, defense, and high-performance engineering fields.

The demands of the aerospace industry to be fulfill by the advanced manufacturing methods and new alloy compositions have been developed to achieve a balanced combination of high strength, improved ductility, toughness, and fatigue resistance. These mechanical behavior improvements also help to reduce the residual stresses in large aircraft components and enhancing its reliability and service life. The integration of advanced ultrasonic inspection methods enables better prediction of fatigue behavior and residual stresses, contributing to safer and more efficient aerospace materials. [5]

Tensile Strength

- Al7068 is recognized as one of the strongest commercially available aluminum alloys, with ultimate tensile strength varies from 220–300 MPa (Without Heat Treated) and even greater than of Al7075. If heat treated varies from 600 to 710 MPa.
- This makes Al7068 stronger than many alloy steels with maintaining lightweight aluminum.

Yield Strength

- Yield strength is the stress level where plastic deformation starts.
- Al7068 provides a very high yield strength of about 140 to 200 MPa (Without Heat Treated).
- This exceptional property ensures superior load-bearing capability before plastic deformation.

Elongation at Break

- Elongation at break indicates ductility or the ability to stretch before fracture.
- Al7068 generally shows elongation varies between 8–12% (Without Heat Treated).
- This improved ductility makes it less brittle.

Modulus of Elasticity

- The modulus of elasticity describes stiffness of a material.
- Al7068 has a modulus of approximately 71–73 GPa as well as similar to Al7075.
- This ensures structural rigidity under applied stresses.

Shear Strength

- Shear strength indicates resistance to sliding forces along a plane.
- Al7068 provides shear strength of about 180 MPa to 250 MPa which is higher than Al7075.
- This makes it ideal for applications such as precision fasteners and load-bearing joints.

Compressive Strength

- Compressive strength measures resistance to forces that push or squeeze material.
- Al7068 typically exhibits compressive strength around 300 - 400 MPa.
- This high compressive strength is useful in aerospace structures and defense components.

Hardness

- Hardness reflects resistance to indentation and wear.
- Al7068 shows a hardness of 50 - 60 HRB (Without Heat Treated), higher than Al7075.
- This provides better wear resistance and surface durability.

Fatigue Strength

- Fatigue strength represents resistance to failure under repeated loading.
- Al7068 demonstrates excellent fatigue strength, superior to Al7075, especially in heat-treated tempers such as T6511.
- This makes it ideal for cyclic loading conditions such as aerospace frames and automotive suspensions.

Fracture Toughness

- Fracture toughness measures resistance to crack initiation and propagation.
- Al7068 has higher fracture toughness compared to Al7075, attributed to its balanced combination of strength and ductility.
- This ensures greater reliability in safety-critical structures like aircraft components and defense systems.

Al7068 not only offers higher tensile, yield, compressive, and shear strength than Al7075, but also provides better ductility, hardness, and fatigue resistance, making it one of the most advanced aluminum alloys available for engineering applications.

One of the strongest commercially available aluminum alloys, Al 7068 frequently outperforms Al7075 in strength-related applications due to its mechanical qualities

1.4 Comparison of Mechanical Properties of Al7075 vs Al7068

Table 1.2 show the mechanical performance of high-grade aluminum alloy Al7075 and Al7068 in terms tensile strength, yield strength, elongation, modulus of elasticity, shear strength, compressive strength hardness, fatigue strength and fracture toughness.

Table 1.2 Comparison of Mechanical Properties of Al7075 and Al7068

Property	Al7075 (Without Heat Treated)	Al7068 (Without Heat Treated)	Key points
Tensile Strength	220–300 MPa	250 – 320 MPa	Al7068 exhibits higher tensile strength, making it the stronger alloy.
Yield Strength	140 – 200 MPa	150 – 220 MPa	Al7068 provides superior resistance to plastic deformation
Elongation at Break	10 – 16%	8 – 12%	Al7068 slightly lower ductility but stronger.
Modulus of Elasticity	69-72 GPa	70-73 GPa	Both alloys exhibit similar stiffness.
Shear Strength	160-230 MPa	180-250 MPa	Al7068 has significantly higher shear resistance.
Compressive Strength	280 – 380 MPa	300 – 400 MPa	Al7068 resists compressive loads more effectively.
Hardness	45 – 60 HRB	50– 60 HRB	Al7068 is harder and more wear-resistant.
Fatigue Strength	Good fatigue resistance	Excellent fatigue resistance	Al7068 outperforms in cyclic loading conditions.
Fracture Toughness	Moderate	Higher than 7075	Al7068 is safer for critical aerospace / defense applications.

1.5 Reinforced Nanocomposites

Reinforced nanocomposites, also referred to as nano-reinforced composites, represent an advanced class of engineered materials where a base matrix is strengthened with nanoscale reinforcements. The combination of nanomaterials imparts superior mechanical, thermal, electrical, and barrier properties that exceed many conventional composites. The following aspects highlight of its significance:

1. **Matrix Materials**
   - The matrix may consist of polymers, metals, ceramics, or hybrid combinations depending on the applications.
   - Polymers such as epoxy, polyethylene, and polypropylene are widely employed due to its ease of processing and versatility.
   - Metals like aluminum and magnesium, along with ceramics such as alumina and silica, are utilized in applications requiring high strength and thermal resistance.
2. **Nanoscale Reinforcements**
   - Reinforcements typically include nanoparticles, nanofibers, or nanotubes.
   - Popular choices include carbon nanotubes (CNTs), graphene, clay nanoparticles, and metal oxides, each imparting distinct functional benefit.
   - Carbon nanotubes and graphene are highly valuable for its exceptional strength, toughness, and electrical conductivity.
3. **Dispersion and Alignment**
   - The effectiveness of a nanocomposite depends on achieving uniform dispersion and orientation of reinforcements within the matrix.
   - Poor dispersion or clustering can degrade performance and introduce defects.
   - Techniques such as sonication, melt mixing, and chemical functionalization are introduced to enhance uniformity and interfacial bonding.
4. **Mechanical Properties**
   - Nanocomposites demonstrate enhanced tensile strength, stiffness, and fracture toughness compared to its base matrices.
   - The nanoscale reinforcements act as barriers to crack propagation, improving structural durability.
5. **Thermal Properties**
   - The addition of high-aspect-ratio reinforcements (e.g., CNTs) significantly increased thermal conductivity.

- It increasing the ability of this nanocomposite to bear excessive heat in aerospace, automotive, and electronics sectors.

6. **Electrical Properties**

- Due to the good electrical conductivity of different reinforcement, nanocomposites can be engineered for highly conductivity.
- CNTs and graphene impart superior electrical conductivity, enable its use in sensors, actuators, and energy devices.

7. **Barrier Properties**

- Nanocomposites often show improved resistance to oxidation, corrosion, and chemical penetration.
- This property enhances its rusting coating ability to resistant against corrosion. Therefore, reinforced nanocomposites offer longer service life and improved reliability in advanced engineering applications.

8. **Applications**

- Its multifunctional nature allows different application across aerospace, automotive, electronics, civil engineering, and biomedical sectors.
- Examples include lightweight structural parts, high-performance coatings, conductive polymers, and controlled drug delivery systems.

9. **Challenges**

- To achieved homogeneous dispersion of nanoparticles still a critical challenge for the researchers, often requiring advanced processing methods.
- Handling and processing of micro or nanoparticles can be posed potential health and environmental risks. Proper safety measures and regulations are necessary.

1.6 Effect of Micro and Nano Reinforced on composites

The incorporation of micro- and nano-scale reinforcements plays a key role in determining the performance and multifunctional behavior of composite materials. These reinforcements, which is available in the form of fibers, particles, or other fillers as well as significantly influence the structural, thermal, electrical, and chemical properties of the composites. The key effects are outlined below:

1. **Strength and Stiffness**

    - Reinforcements greatly enhance the mechanical strength as well as stiffness of composites.
    - By effectively gradually applied the applied load throughout the matrix, it enhances tensile, compressive, and flexural properties.
    - The modulus of elasticity is also increased as well as imparting higher rigidity and resistance to deformation.

2. **Toughness and Impact Resistance**

    - Reinforcements can improve toughness and increase impact resistance.
    - The aluminum alloy composite can absorb and dissipate energy under dynamic loading as well as reducing the risk of sudden or impact-related failure.
    - This property is especially important for components subjected to shocks, impacts as well as repeated cyclic loading.

3. **Weight Reduction**

    - Micro- and nano-reinforcements having important role in reducing the overall weight of composite materials as well as maintaining or even improving its mechanical performance.
    - Micro-reinforcements also contribute to weight reduction by increasing the hardness and strength as well as replaced the heavier conventional materials such as steel with lighter aluminium-based composites.

4. **Thermal Conductivity**

    - Nanomaterials such as carbon nanotubes (CNTs) as well as graphene can substantially improve the thermal conductivity of composites.
    - This makes reinforced composites is suitable for thermal applications, including heat sinks as well as electronic devices.

5. **Electrical Conductivity**

- Some reinforcements can improve electrical conductivity to the composite.
- Carbon-based materials such as CNTs and graphene are especially useful in applications like electronics, sensors, as well as electromagnetic interference (EMI) shielding

6. **Dimensional Stability**

- Reinforcements reduce the coefficient of thermal expansion (CTE) as well as induce greater dimensional stability under variable temperature.
- The proper dimensions is important for precision-engineered components where dimensional accuracy must be necessary.

7. **Wear and Abrasion Resistance**

- The mixing of micro-sized particles into aluminium alloys improves the load-bearing capacity also as well as reduces material loss during sliding or abrasive conditions.
- Nano-reinforcements provide large surface area also as well as providing interfacial bonding with the metal matrix composite, which helps in improving hardness, strength, and resistance to surface damage.

8. **Chemical Resistance**

- Micro- and nano-particles increases the chemical resistance ability of composite materials as well as acting as protective barriers against corrosive environments.
- Silicon carbide, alumina, graphene, and other micro and nano-particles possess high chemical stability as well as offer resistance to oxidation or corrosion.

9. Fatigue Resistance

- Reinforcements can improve the fatigue life by slowing crack initiation and growth.
- Nano-reinforcements (e.g., CNTs, graphene) are especially effective in bridging micro-cracks as well as delaying failure.

10. Creep Resistance

- Reinforcements can also enhance the material's ability to resist long time deformation under constant stress at elevated temperatures.
- Ceramic nanoparticles (like SiC, Al_2O_3) are having useful for improving creep resistance in metal matrix composites.

11. Fracture Toughness

- Nano-reinforcements increase crack deflection and energy absorption.
- Micro-scale reinforcements also improve fracture resistance but it's may less effective at the nanoscale crack-arresting level.

12. Thermal Stability

- Nanoparticles enhance thermal stability by restricting polymer chain mobility or grain growth in metals.
- This ensures performance retention at high temperatures.

13. Damping Properties

- Reinforcements can enhance vibration damping and energy dissipation.
- This is valuable in aerospace and automotive structures for noise and vibration reduction.

14. **Challenges**

- The performance of reinforced composites depends on the type, size, distribution, and alignment of reinforcements.
- Achieving uniform dispersion, particularly at the nanoscale, is technically challenging and often requires advanced processing methods.
- Balancing performance enhancements with cost and manufacturability remains a crucial consideration in composite development.

1.7 Reinforcement of micro particles with Silicon Carbide

The mixing of micro-sized silicon carbide (SiC) particles into a base material is a well-established process to improve the overall performance of composites. Silicon carbide is a ceramic material recognized for its extreme hardness, high thermal conductivity, and excellent chemical stability. When used as micro-scale reinforcements within a matrix, SiC imparts significant improvements to mechanical, thermal, and wear properties of the composite. The effects are summarized below:

1. Mechanical Properties

- Micro-sized SiC particles strengthing the aluminium alloy by removing dislocation motion as well as improving load transfer.
- As a result, clear improvements can be seen in tensile, compressive, and flexural strength is happened, making the material suitable for high-stress structural applications.

2. Toughness and Impact Resistance

- The mixing of SiC micro-particles can enhance the composite's toughness by removing crack propagation.
- This improves resistance to sudden impact as well as reduces the chances of brittle fracture under service conditions.

3. Wear and Abrasion Resistance

- SiC particles provide excellent wear as well as abrasion resistance.
- Composites reinforced with SiC can withstand prolonged period of time with friction and erosion as well as making it ideal for components such as brake discs, seals, and cutting tools.

4. Thermal Stability

- SiC is having very high thermal conductivity and stability.
- The reinforcement allows the composite to perform well under high temperatures as well as resistance to thermal fatigue and thermal shock.

5. Dimensional Stability

- The mixing of SiC micro-particles lowered the coefficient of thermal expansion (CTE).

- The SiC micro-particles minimize the thermal distortion as well as enhances dimensional precision during fluctuating temperature cycles.

6. **Electrical Properties**

- Silicon carbide is having a semiconductor can modify the electrical behavior of composites.
- It can also be used to either enhance conductivity or provide controlled insulating properties in electronic as well as structural systems.

7. **Chemical and Oxidation Resistance**

- SiC is having strong resistance against oxidation and chemical attack.
- This makes SiC-alumnium alloy composites is particularly suitable against aggressive chemical environments as well as at high-temperature industrial processes.

8. **Challenges**

- Achieving uniform dispersion of micro-sized SiC particles within the matrix is a persistent challenge, as agglomeration can lead to stress concentration sites.
- Processing methods such as powder metallurgy, stir casting, or hot pressing are required to optimize distribution.
- Additionally, the extreme hardness of SiC can complicate machining and finishing operations.

1.8 Reinforcement of micro particles with Silicon Nitride (Si_3N_4)

Silicon nitride (Si_3N_4) is a high-performance ceramic material widely valued for its superior mechanical strength, low density, and excellent resistance to wear, thermal shock, and chemical degradation. When incorporated into a matrix material as reinforcing particles (micro or nano scale), Si_3N_4 imparts substantial improvements to structural, thermal, and functional properties of composites. Its role as reinforcement is summarized below:

1. **Mechanical Properties**

- Si_3N_4 particles significantly enhance the load-bearing capacity of the matrix.
- Its intrinsic hardness and high modulus contribute to improved tensile, compressive, and flexural strength.
- Crack deflection and particle–matrix interactions help in resisting mechanical failure under high stresses.

2. **Toughness and Impact Resistance**

- Unlike many brittle ceramics, Si_3N_4 exhibits relatively high fracture toughness.
- When used as reinforcement, it increases the impact resistance of the composite and delays crack initiation and propagation.

3. **Wear and Friction Resistance**

- Silicon nitride particles impart excellent wear and frictional resistance due to its extreme hardness.
- Components reinforced with Si_3N_4 show longer lifespans in abrasive or sliding-contact environments, such as bearings, seals, and engine components.

4. **Thermal Stability and Shock Resistance**

- Si_3N_4 having outstanding thermal shock resistance compared to many other ceramics.
- Reinforced composites having endure rapid heating and cooling cycles without cracking as well as making it suitable for aerospace and automotive engine parts.

5. **Dimensional Stability**

- Mixing of Si_3N_4 reduces the coefficient of thermal expansion (CTE) as well as enhancing dimensional accuracy in fluctuating thermal conditions.
- This property is important in precision engineering work as well as for electronic packaging applications.

6. Electrical Properties

- Silicon nitride is an electrical insulator with excellent dielectric strength.
- Reinforced composites can also maintain or enhance electrical insulation as well as it is useful in electronic and high-voltage applications.

7. Chemical and Oxidation Resistance

- Si_3N_4 is having chemically inert and maintains its integrity in corrosive and high-temperature oxidative environments.
- This makes reinforced composites is highly durable in aggressive industrial as well as in chemical processing applications.

8. Lightweight Nature

- With relatively low density compared to other ceramics, Si_3N_4 is helpful to reducing the overall weight of composites.
- It is use in aerospace, defense, and transportation sectors where weight savings are critical.

9. Challenges

- Achieving uniform distribution of Si_3N_4 particles in the matrix can be challenge for researcher due to particle agglomeration.
- Strong bonding between the reinforcement and the aluminium alloy is necessary to get the desired properties. Methods such as stir casting, powder metallurgy, and hot isostatic pressing are commonly used.
- Due to the high cost of high-purity Si_3N_4 powder, can't use in excess in large-scale applications.

1.9 Micro SiC (Silicon Carbide) Particles

Micro silicon carbide (SiC) particles are broadly used as reinforcements in aluminium alloy due to having excellent mechanical, thermal, and wear-resistant properties. SiC is also having hard micro particle known for its lightweight nature, excellent stability, and ability to perform under extreme conditions. Here are the detailed effects and applications of micro SiC particles:

1. **Mechanical Reinforcement**

- Micro SiC particles, having high strength and stiffness and hence significantly improve the strength, stiffness, and hardness of metal, polymer, and ceramic matrices.
- Micro SiC particles, having a high modulus of elasticity as well as provide better load-bearing capacity and resistance to deformation
- Composites reinforced with micro SiC particles, having improved mechanical properties, are useful in lightweight aerospace as well as in automotive applications.

2. **Wear Resistance**

- SiC is having higher hardness as well as enhances the wear and abrasion resistance of composites.
- Aluminium alloy reinforced with micro SiC particles as well as having improved resistance to wear & friction and show longer service life in abrasive environments.
- These composites, having high durability and wear resistance. So it is widely use in brake discs, bearings, seals, and cutting tools.

3. **Thermal Properties**

- SiC is having very high thermal conductivity, improves heat dissipation in composites.
- SiC is having excellent thermal stability as well as it is suitable for applications exposed to high temperatures and thermal cycling.
- This aluminium alloy is having good heat resistance and is widely used in heat exchangers, engine components, and high-temperature molds.

4. **Electrical Properties**

- SiC is having semiconducting properties and can introduce controlled electrical conductivity into composites.
- This aluminum alloy reinforced with SiC is having improved electrical behavior, are useful in electronic devices, sensors, and electromagnetic shielding applications.

5. **Corrosion and Oxidation Resistance**

- Micro SiC particles is having strong resistance to oxidation and corrosion as well as enhance the durability of composites.
- These composites are having good chemical and thermal stability as well as it is widely used in harsh chemical and high-temperature environments.

6. **Lightweight Advantage**

- SiC is having low density compared to metals as well as helps to reduce weight while maintaining mechanical strength.
- These materials, having a high strength-to-weight ratio, are critical for aerospace and defense applications.

7. **Challenges**

- Micro SiC particles is having a tendency to agglomerate, make it difficult to achieve uniform dispersion within the matrix.
- Processing methods, having the capability to ensure better distribution, such as advanced techniques such as powder metallurgy and liquid infiltration.

1.10 Micro Si_3N_4 (Silicon Nitride) Particles

Micro silicon nitride (Si_3N_4) particles are high-performance ceramic reinforcements and it's having unique balance of strength, toughness as well as thermal shock resistance. These are widely using in advanced engineering fields where extreme durability is required.

1. **Mechanical Reinforcement**

- Micro Si_3N_4 particles is having enhance tensile and flexural strength as well as providing superior fracture toughness compared to most ceramics.
- It can resist crack initiation and propagation as well as increasing composite durability.
- Used in structural components, engine parts, and load-bearing applications.

2. Wear Resistance

- Si_3N_4 is having highly resistant to wear, abrasion as well as rolling/sliding contact fatigue.
- It is frequently uses in ball bearings, cutting tools, and mechanical seals.

3. Thermal Properties

- Si_3N_4 is having highly resistant against thermal stress and making it right for quick heating and cooling.
- Its stability at high temperatures makes it ideal for turbine blades, engine valves, and heat-resistant coatings.

4. Electrical Properties

- Si_3N_4 is having an excellent electrical insulator with high dielectric strength.
- It is broadly used in electronic instruments, insulating components, and high-voltage applications.

5. Corrosion and Oxidation Resistance

- It is highly resistant to oxidation and corrosion caused by strong chemicals.
- This makes it suitable for use in chemically harsh or high-temperature industrial environments.

6. Lightweight Advantage

- Si_3N_4 offers a high strength-to-weight ratio, making it desirable in aerospace and defense systems where both lightness and toughness are required.

7. Challenges

- Processing of micro-Si_3N_4 particles into metal matrix composites is challenging and sometimes requires advanced sintering or hot isostatic pressing (HIP).
- Higher cost compared to conventional reinforcements limits its large-scale industrial adoption.
- Its relatively high cost compared to conventional reinforcements restricts mass industrial applications.

Chapter 1 : Introduction

Now these days materials are being meticulously designed and engineered to tackle challenges in aerospace engineering and exploration. At the same time, materials science, especially nanomaterials science, is flourishing as a field of research. Scientists are now creating advanced materials with outstanding functionalities and, in many instances, multifunctional capabilities.[1] Currently, aluminum matrix composites are in high demand in the aerospace industry, automobile industry, and other engineering applications.[2] In this context, lightweight materials such as composites are formed by combining two or more different materials to achieve enhanced physical and mechanical properties compared to the individual materials, fulfills the needs of various industries and modern engineering applications.[3]

Aluminum-based alloys are ideal for aircraft structure because of their ease of fabrication, outstanding thermal conductivity, cost-effectiveness, and recyclability, which align with sustainability goals. The properties of aluminum can be greatly enhanced through the careful addition of suitable reinforcements.[4] Known for being relatively lightweight and having excellent thermal and electrical conductivity, aluminum alloys can be further customized to meet specific engineering needs by incorporating various reinforcements. These additions, such as nanoparticles, microparticles, and fibers, can significantly enhance the material's properties, including tensile strength, wear resistance, and porosity. Researchers frequently incorporating nano particles or micro particles such as MgO, SiO_2 , Al_2O_3, SiC, TiC, Si_3N_4 , BN, TiB_2 , graphite may enhance the tensile strength, hardness, wear rate and yield strength of the composite material.[5][6]

Furthermore, the study revealed a significant improvement in hardness and tensile strength with an increase in the weight percentage of Silicon Carbide in the composites. Microstructural analysis confirmed a uniform distribution of Silicon Carbide (SiC) and Silicon Nitride (Si_3N_4) reinforcements within the as-cast Al-7068 metal matrix composite. The addition of reinforcement particles (in varying wt%) to Al-7068 significantly improved the hardness and tensile strength. The composites were successfully fabricated using the stir casting technique.[6]

1.11 Organization

Chapter 1: Introduction: This chapter highlights the use of aluminum alloys in modern engineering, with specific focus to Al7068 for its remarkable strength-to-weight ratio. It studies the foundation underlying strengthening these alloys using micro-sized SiC and Si_3N_4 particles, aiming to enhance important mechanical properties such as hardness, tensile strength, and wear resistance. By framing these inspirations, this chapter launches the basis for the research, clearly defining its objectives and delineating the scope of the investigation.

Chapter 2: Literature Review: This chapter gives a comprehensive review of existing research on the reinforcement of aluminum alloys with nano and micro particles. It explains the various methods adopted to mixed SiC and Si_3N_4 particles into metal matrix composite and evaluates the effects on its mechanical properties. The main review is on the comparative studies that distinguish between nano- and micro-sized reinforcements as well as helping for selecting specific particle sizes. Through this analysis, the chapter identify critical gaps in previous research work and establishes the set up for the current research objectives pursued in the thesis.

Chapter 3: Materials and Experimental Details: This chapter presents a detailed description of the materials selection and fabrication of casting specimen for the investigation on the Al7068 alloy and the mixing of SiC and Si_3N_4 particles as reinforcements. The fabrication of the composite specimens is carried out using stir casting techniques, ensuring proper distribution of the SiC and Si_3N_4 particulates as per composition decided. In addition, the chapter outlines the experimental testing perform at Material Testing Laboratory in MANIT, Bhopal to evaluate the mechanical properties of the developed composites. Standard methods are used for Rockwell hardness testing, Tensile strength, Yield strength, Elongation and Toughness measurement, with careful attention to reproducibility and accuracy in order to validate the reliability of the results.

Chapter 4: Results and Discussion: This chapter presents a detailed description of the experimental outcomes such as Tensile test, Hardness test, Toughness test and highlighting the influence of micro-sized SiC and Si_3N_4 particles on the mechanical performance of Al7068 alloys. The results are systematically analyzed with the help of FESEM & Optical

Microscopic images and identify the optimal particle size and distribution within the metal matrix, after this find the conditions that yield the most significant improvements. After this discussion further explores the physics behind the observed enhancements and compares these findings with the hypotheses and other studies cited in the literature review.

Chapter 5: Conclusion and Future Scope: The final chapter synthesizes the final results of the study, shown that the mixing of micro-sized SiC and Si_3N_4 particles by stir casting process is notable enhancements of the mechanical behavior of Al7068 alloys. These reinforcements significantly improve hardness, tensile strength, and impact strength as well as validating the effectiveness of SiC and Si_3N_4 particles additions in high-performance Al7068 aluminum metal composites. This chapter gives the conclusion by offering future research work, highlighting the scope to investigate alternative micro particles and to perform mechanical experimental test at normal environmental conditions. Such research work will expand the potential applications of reinforced Al7068 alloys and contribute to the advancement of lightweight, high-strength materials in engineering practice.

1.12 Chapter Summary

The introductory chapter of this thesis gives the important role of Al7068, a widely used aluminum alloy known for its excellent strength-to-weight ratio, in various engineering applications. It highlights the innovative approach of reinforcing this alloy with micro- sized silicon carbide and silicon nitride particles to potentially enhance its mechanical properties, including hardness, tensile strength, and wear resistance. This improvement is aimed at extending the alloy's applicability in more demanding industrial environments.

The chapter also presents the objectives of the study, which focus on comparing the effects of silicon carbide and silicon nitride reinforcements, and establishes the research methodology that will be employed to assess these effects. Through this investigation, the thesis aims to contribute valuable insights to the field of materials science and engineering, particularly in the development and optimization of high-performance materials.

Chapter - 2
Literature Review

Chapter 2
Literature Review

2.1 Introduction

The literature review chapter also examines the existing studies on the reinforcement of Al6061, Al7075 & Al7068 alloy with various reinforcements such as ZrO2, silicon carbide (SiC) and silicon nitride (Si_3N_4). These micro particles are widely being used to increase the hardness, thermal stability, and wear resistance as well as making it best alternative for improving the mechanical properties of high-strength aluminum alloys metal matrix composite.

The literature review is based on research conducted over the past 20 years, with a primary focus on the influence of various nano and microparticles on the mechanical properties of aluminum alloys, as well as the associated fabrication processes.

Comparative analyses in the literature show that effect of particle size and distribution of these reinforcements. Micro-sized particles have mechanical properties to improve bulk strength and stiffness, while nano-sized reinforcements have characteristics to grain refinement and enhanced fatigue resistance. However, uniform dispersion of these ceramics' particles composition continues to be a substantial challenge, that affects the overall performance of the composite.

By analyzing previous research work, this chapter recognizes key research gaps, such as limited work have been done to optimizing hybrid reinforcements (SiC + Si_3N_4) in Al7068 and scarcity of investigations performed into the long-term durability of these composites under extreme service conditions. These gaps set up the framework for the present study, which goals to offer a more thorough investigation into how ceramic reinforcement influences the mechanical, thermal, and wear characteristics of Al7068 alloys

2.2 Literature Review on Mechanical Properties & Fabrication methods

The literature review has been done on based of influence of various nano and microparticles on the mechanical properties of aluminum alloys and its fabrication process.

The researcher showed the importance of aluminum matrix composites alloy by using in the application of automobile and aerospace. Aluminum alloy reinforced with different micro-particles like carbides, nitrides, oxides, and borides offer improved mechanical strength and wear resistance as compared to other conventional materials. Author emphasizes the stir casting process due to its cost-effective and scalable manufacturing technique as well as discussing about its process parameters, advantages and challenges. These study summarizes key property enhancement mechanisms, current applications, and future research directions for aluminium matrix composites.[6]

This review highlights effects of corrosion on the aluminium alloys and aluminium-based metal matrix composites widely used in industrial applications. Although these materials provided excellent strength-to-weight ratio and inherent corrosion resistance. The study discusses how different reinforcements influence the corrosion behavior of aluminium hybrid composites, along with the manufacturing methods of casting, corrosion testing methods, and surface characterization approaches. Special attention is given to corrosion inhibitors as an effective strategy for improving durability and service life.[7]

This study investigates the impact of graphite (Gr) and tungsten carbide (WC) reinforcements on the mechanical properties of Al6061 alloy composites manufactured by using the stir-casting method. The results show that increasing tungsten carbide (up to 6%) and graphite (up to 4%) enhanced the tensile strength, yield strength, hardness, and stiffness properties. The best results were achieved with percentage composition of 4% Gr and 6% tungsten carbide, which showed notable improvements over the base alloy. A deep learning–based DLSTM model was successfully used to predict mechanical properties with high accuracy. The findings demonstrate the potential of WC–Gr reinforced Al6061 composites for developing lightweight, high-strength materials for automotive and aerospace applications[8]

This review summarizes microstructural, mechanical, and tribological studies carried out under varying reinforcement content and wear conditions. Statistical tools such as Taguchi methods and ANOVA reveal that applied load strongly influences wear behavior, while sliding speed plays a key role in reducing friction and material loss. Overall, uniform reinforcement distribution and controlled testing conditions are crucial for enhancing the performance of AMMCs.[9]

This study investigates that how added graphene to aluminium-based metal matrix composites affects the field emission performance, which is critical for advanced electronic and energy applications. Due to experimental limitations at higher graphene contents, a two-stage machine learning framework was developed to predict emission current density and stability for graphene contents between 1.25 and 2 wt%. Author found that the composite with mixing of 2 wt% graphene exhibits the best field emission performance, generate optimal graphene dispersion and conductive network formation within the aluminium matrix. This study highlights the effectiveness of machine learning is a powerful tool for optimizing composite materials for high-performance applications and supervisory future experimental research.[10]

This review explains the machining process of aluminium metal matrix composites (AMMCs) by using electrical discharge machining (EDM), also highlighting both its capability and challenges. It explains the influence of the EDM process by the micro particle's composition and properties of AMMCs as well as discussed about key benefits such as precise machining of complex shapes and effective handling of materials with varying hardness. The review also showing challenges of electrode wear, surface quality issues, and the effect of reinforcement particles on machining efficiency. Emphasis is placed on optimizing EDM parameters to improve material removal rate and surface finish, while also considering environmentally sustainable machining practices.[11]

This review highlights the developing interest in the research area of aluminium metal matrix composites due to the need for lightweight, strong, and efficient materials across industries such as automotive, aerospace, marine, and defense. It focuses on hybrid aluminium composites produced using the stir casting method due to its simplicity, low cost, and suitability for large-scale production. The paper discussed about uses of the aluminium alloys, reinforcement materials, and its influence on microstructural, mechanical, and tribological properties. The review serves as a practical guide for researchers in selecting suitable matrix–reinforcement combinations for specific engineering applications.[12]

This review focuses on aluminum alloy nano-composites material, highlighting the mixing of micro- and nano-particles reinforcements improves the mechanical properties of aluminum alloys. The main research has been done on Al7075 aluminum alloy composites due to its high strength, good wear and corrosion resistance, and thermal stability. The paper discussed about the use of various microparticles and nanoparticles such as Al_2O_3, SiC, TiO_2, ZrO_2, graphite, and TiB_2, and reinforced with heat-treatable alloys like Al2024, Al6061, and Al7075 are mostly preferred.[13]

This study focuses on aluminum hybrid composites, which use multiple reinforcements to achieve better mechanical and tribological performance instead of single-reinforced materials. Electric Discharge Machining (EDM) are used to machine such hard composites. The composites have been manufacturing with the help of liquid stir casting method, and machining parameters has been optimized by using the Taguchi method with an L27 orthogonal array. The main objective of this research is to improve the material removal rate and identify the optimal EDM machining conditions. The results were validated through confirmatory experiments, demonstrating the effectiveness of the selected parameters for machining advanced aluminum hybrid composites.[14]

This paper provides a detailed review of previous research on nanoparticle-reinforced aluminium matrix nanocomposites (AMNCs). Metal Matrix Composites (MMCs) play an important role in aerospace, defense, and automotive industries due to its superior performance. The survey focuses on the mechanical behavior and surface characteristics of composites reinforced with nanoparticles like MgO, SiO_2, Al_2O_3, SiC, TiC, BN, Si_3N_4, and TiB_2, and also discusses the various fabrication techniques of MMCs.[15]

Most studies have concentrated on aluminium alloys like 356, 2024, 6061, and 7075, owing of its corrosion resistance, wear resistance, thermal stability, annealing capability, and ease of casting. The incorporation of nanoparticles has been shown to significantly improve tensile strength, hardness, yield strength, and wear resistance of aluminium-based composites.[15]

This study focuses on machining of Aluminum 7075 metal matrix composites optimizing by abrasive water jet machining (AWJM). Using Taguchi and DEAR methods, the effects of pressure, standoff distance, and traverse speed on material removal rate, surface roughness, and taper angle were analyzed. The optimal machining condition was identified as 180 MPa pressure, 3 mm standoff distance, and 40 mm/min traverse speed, which resulted in noticeable improvements in surface quality. Traverse speed was found to be the most influential parameter. SEM analysis of the machined surface revealed typical features such as ploughing marks, ductile fracture, and minor cracking, confirming the influence of machining conditions on surface integrity.[16]

This study shows the tribological improvement of Al7075-SiC composites through the mixing of graphite as a solid lubricant. While Al7075-SiC composites are widely used in engineering applications, due to its friction and wear characteristics limitations. For this purpose, Al7075-SiC/Graphite hybrid composites containing 3–9 wt% graphite have been fabricated via stir casting and tested under dry sliding conditions.[17]

Using Taguchi optimization, ANOVA, and regression analyses, the influence of sliding distance, load, and graphite content on wear and friction was systematically examined. The results indicate that 9 wt% graphite reinforcement noticeably decreases both wear rate and coefficient of friction, made the self-lubricating composites. From the among the parameters, sliding distance dominantly affected wear behavior, whereas graphite composition (77% contribution) applied the extreme influence on frictional response, with load is a secondary role. Overall, the work demonstrates that graphite addition effectively transforms Al7075-SiC composites into advanced self-lubricating materials, thereby broadening its applicability in demanding tribological environments.[17]

This work investigated the microstructural changes and mechanical performance of Al–10Zn–3.5Mg–2.5Cu/graphene nanoplate (GNP) composites prepared by via ball milling and stir casting. Reinforcement of GNPs, due to its large specific surface area, significantly improved the strength characteristics of the composites. Microstructural observations indicated that GNP addition improved the grain refinement, although at higher content GNPs (1 wt%) no further reduction in grain size was observed.[18]

Tensile testing demonstrated that GNP reinforcement has been enhanced ultimate tensile strength (UTS), with the addition of 0.7 wt% GNP composite and achieving UTS 582 MPa. Furthermore, T6 heat treatment (solutionizing at 460 °C for 8 h, water quenching, and aging at 120 °C for 24 h) resulted the superior strength compared to other processing conditions. However, excess amount of GNP addition led to agglomeration at grain boundaries, which acted as preferential sites for crack initiation and propagation. Overall, the conclude that a controlled addition of GNPs—particularly at 0.7 wt%—optimally enhances the strength of Al–Zn–Mg–Cu alloys, while excessive reinforcement can decreases fracture resistance due to agglomerate formation.[18]

This review discusses about the importance of ultra-high strength steels in aerospace and defense applications and highlights the challenges for improving the mechanical properties such as strength and toughness. It also summarizes three major types of martensitic ultra-high strength steels and explains in details how all these mechanical properties can be improved through advanced strengthening and toughening strategies. The study emphasizes that combining nano-scale precipitates with highly dislocated martensitic structure is a best procedure for developing next-generation ultra-high strength steels with superior overall performance.[19]

This study investigated the lightweight high-entropy alloys (LHEAs) used as reinforcements for Al-7068 aerospace alloy composites. LHEA composition (Al–Li–Mg–Be–Ti–Cu) have been added at 4, 10, and 16 wt% using stir casting, with the aim of improving both strength and ductility. X-ray diffraction analysis report confirmed the formation of BCC and FCC phases originating from the LHEA, along with secondary $CuAl_2$ and $MgBe_{13}$ intermetallic phases. The microstructural distribution of LHEA particulates varied with different amount content at : sparse at 4%, uniform at 10%, and partially clustered at 16%.[20]

This review discusses about the use of nanoparticle-reinforced aluminium metal matrix composites to meet the application of aerospace industry's demand for lightweight yet high-performance materials. While aluminium and its alloys are attractive due to its low weight and cost, its limited wear resistance and thermal stability restrict wider application. This study exploring the impact of various nanoparticles to improve the mechanical strength and wear behaviour of aluminium composites, with focus on stir casting manufacturing method. It also highlights how the different mechanical properties of these composites can be developed for specific applications, along with the key challenges associated with nanoparticle reinforcement in aluminium MMCs.[21]

This review analysis explores the mechanical performance of aluminium matrix composites reinforced with silicon carbide, granite, and graphite powders. The composites have been manufactured with help of the stir casting technique by varying the reinforcement proportions and investigations have been done according to ASTM standards for hardness, tensile strength, and shear strength. The results show notable improvements in strength and wear resistance compared to conventional materials, highlighting the suitability of these composites for applications that require high durability and abrasion resistance.[22]

This study compared the mechanical properties and microstructure of copper-based hybrid composites alloys reinforced with SiC–graphite and WC–graphite microparticles, produced by using the stir casting method. The results show that both microparticles are uniformly distributed throughout the copper matrix and are clearly identified through microstructural and phase analyses. With the mixing of hard ceramic microparticles significantly improves the tensile strength and hardness in both composites. Among them, WC–graphite reinforced composites exhibited superior mechanical properties compared to SiC–graphite reinforced ones. Fracture analysis indicates that both composites firstly fail through ductile fracture test, confirming its ability to retain toughness while gaining strength.[23]

This study developed an eco-friendly aluminum composite by reinforcing with Al6061-T6 alloy with waste-derived materials such as palm kernel shell ash, rice husk ash, and steel particles. Using response surface methodology to analyzed the effects of reinforcement content and stirring temperature on microstructure and strength. Results showed that moderate amounts of ash particles significantly improved tensile, flexural, and compressive properties, while excessive mixing of ash particles reduced strength. Optimal performance was achieved at a controlled reinforcement ratio and stirring temperature, and the experimental results closely matched the predicted models. The study demonstrates the strong potential of waste-based hybrid reinforcements for sustainable automotive aluminum composites.[24]

This research paper reviews the different types of aluminium matrix nanocomposites (AMNCs), highlighting superior structural and mechanical properties compared to its individual constituents. The focus is on fabrication method of Al7075 based composites as well as on its mechanical properties study such as high strength, corrosion resistance, wear resistance, and thermal stability, are widely employed in aerospace, automotive, defense, marine, and industrial sectors.[25]

Particular attention is given to heat-treatable alloys (Al2024, Al6061, Al7075) as well as on its processing ease, heat-treatability, and adaptability for tailored applications. This work examined the addition of various micro- and nano-reinforcements, including Al_2O_3, ZrO_2, TiO_2, SiC, graphite, and TiB_2, and its role to enhancing mechanical properties such as tensile strength, hardness, wear resistance, and thermal stability. The research findings demonstrating that the combination of ceramic and carbon-based reinforcements into Al7075 and other aluminium alloys considerably improves mechanical performance, thereby extending its application in high-demand engineering field.[25]

This research explores the impact of zirconium dioxide (ZrO_2) reinforcement on the mechanical performance of Al7075 alloy, with focus to make compositions for aerospace-grade applications. Al7075–ZrO_2 composites has been synthesized via stir casting, mixed with both micron-sized (1–3 wt%) and nano-sized (0.17–0.51 wt%) ZrO_2 particles. Mechanical testing evaluated peak stress, load capacity, and yield strength.[26]

Results revealed that nano-ZrO_2 reinforcement has more effective than micron-scale additions. The 0.51 wt% nano-ZrO_2 composite exhibited the highest peak stress and load, while 0.34 wt% nano-ZrO_2 achieved a 59% increase in peak stress (153 MPa) and a 24% reduction in strain (3.7%) compared to pure Al7075 (125 MPa, 4.9%). In contrast, the micro-ZrO_2 composite measured 125 MPa peak stress with 2.9% strain. The findings show that nano-ZrO_2 enhanced the load-bearing capacity and tensile strength, making Al7075–ZrO_2 composites highly suitable for aerospace applications demanding excellent resistance to tensile failure.[26]

In this research paper, author review and did deep study in the field of the increasing demand for lightweight, energy-efficient, and environmentally sustainable materials has intensified research on aluminum-based metal matrix composites (AMMCs), particularly made from the 7xxx series alloys. Due to its superior strength-to-weight ratio, aluminum alloys serve as an excellent base raw material for composites and its properties can be further enhanced through nanoparticle reinforcements. Researches surveys latest researches in the area of Al7075-based MMCs reinforced by taken both organic and inorganic nanoparticles fabricated via diverse processing routes, with particular on its mechanical properties and tribological performance. From the various fabrication methods, stir casting with ceramic nanoparticles has been shown the most consistent improvement in strength, hardness, and wear resistance, making it especially perfect for aerospace applications. Finally, the reviewer identified the challenges—such as nanoparticle agglomeration and interfacial bonding—and outlines future research pathways for optimizing fabrication methods and achieving superior property tailoring.[27]

Studied the effect of MoS_2, SiC and B_4C by varying wt % (3,6,9 & 12%) on the hybrid Al6061 metal matrix composite and harness has been improved by increasing content of B_4C. Maximum hardness 114.03 HV showed at 12% B_4C/ 4% MoS_2.[28]

The author demonstrates that hybrid aluminum composites reinforced with TiB_2 and graphene can improve the wear resistance, meeting the increasing demand for durable engineering materials. Increasing graphene content effectively reduces wear rate, highlighting its strong lubricating and load-bearing role. The wear behavior is strongly influenced by temperature, shifting from adhesion-dominated wear at room temperature to abrasion-driven wear at elevated temperatures, with severity increasing as temperature rises. Microstructural and phase analyses confirm uniform reinforcement distribution and explain the observed wear mechanisms. Overall, the developed hybrid composite shows strong potential for high-temperature and wear-critical applications. [29]

This study examined the isothermal hot compression properties of Al-7.3Zn-2.2Mg-2Cu (Al7068) alloy at T6 condition up to true strain at 0.69, within the temperature range of 250–450 °C and strain rates of 0.001–1

s^{-1}. The rheological response indicated that clear reduction in flow stress at higher temperatures and lower strain rates, with the apparent activation energy for deformation determined as 266 kJ/mol.[30]

The 3D processing map identified two optimum hot-working domains range: Zone I (288–312 °C, 0.001 s^{-1}) with a peak efficiency of 57%, and Zone II (400–450 °C, 0.001–0.044 s^{-1}) with an efficiency of above 48%. Overall, the work provides a comprehensive understanding of the hot deformation mechanisms of Al7068, showing processing method and microstructural changes pathways critical for aerospace applications.[30]

The study shows that reinforcing Al1050 with $Al_{0.5}CoCrFeNi$ high-entropy alloy particles improved its mechanical performance. The addition of 3 wt% HEAp has increased the tensile strength compared to the base alloy. Among the post-processing techniques, cryorolling proved that more effective method than room-temperature rolling, resulting in superior strength and microstructural refinement. Cryorolling also suppressed the formation of microvoids, which were commonly observed in room-temperature rolled composites. Inclusive, the combined effect of HEA reinforcement and cryogenic deformation enhanced both the strength and structural integrity of Al1050-based MMCs, highlighting cryorolling is best manufacturing method for high-performance composite fabrication.[31]

This research investigated the fabrication and mechanical performance of Al7075/TiO_2 metal matrix composites (MMCs) manufactured via stir casting method with TiO_2 micro particles additions varies from 2–10 wt%. The comprehensive characterization study has been conducted, such as tensile testing, hardness, dynamic mechanical analysis, XRD, and microstructural evaluation, alongside tribological tests using a high-temperature pin-on-disc setup against EN31 steel.[32]

Microstructural analysis report confirmed that uniform TiO_2 dispersion, grain refinement, and phase evolution, indicated improvements in mechanical performance. Tensile testing report showed that progressive increases in ultimate strength and load capacity with the increase of TiO_2 micro particles content, while hardness measurements (Vickers hardness test) consistently indicated enhanced mechanical resistance, underscoring the alloy's potential for high-strength and wear-critical applications. Dynamic mechanical test report exposed that superior damping and flexural response at 8 wt% TiO_2, with elastic modulus value varies from 98.9 GPa (25 °C) to 77.3 GPa (350 °C). Tribological properties of developed material Al7075 reinforced with TiO_2 presenting lower coefficients of friction and wear rates, wear behavior on temperature-dependent influenced by localized heating and micro-welding. Overall conclude that the addition of TiO_2, mainly at 8 wt%, optimizes the mechanical, thermal, and tribological properties of Al7075, which is suitable for structural and high-temperature engineering applications.[32]

This research highlights the increasing demand of aluminum alloy metal composites in the applications of aerospace and automotive because it provides high strength-to-weight ratio and thermal stability. It discusses the different types of microparticles and nanoparticles influence on the mechanical properties of these composites and reviews recent development in this field. Researchers also outlines the future opportunities and perspectives for the development and application of aluminum alloys composites in engineering structures.[33]

This study discusses the development of a hybrid Al7075 metal matrix composite (MMC) reinforced with both sustainable and synthetic particles, produced by using a two-stage stir casting process assisted by ultrasonic transduction. The synthesized composites have been subjected to mechanical, tribological, and microstructural analysis. Microstructural analysis via optical microscopy and FESEM with elemental mapping confirmed a uniform dispersion of reinforcements across the aluminium matrix. [34]

The addition of reinforcements greatly improved the compressive strength and microhardness, with the B3 composite attaining peak values of 569.17 MPa and 178.86 HV, respectively, compared to the as-cast Al7075 alloy. Tribological analysis reported that the wear resistance improved with mixing of reinforcement, with the B3 sample showing the lowest wear rate. FESEM analysis of worn surfaces revealed grooves, micro-cracks, pits, and wear debris, indicating abrasion and delamination driven wear mechanisms. Overall, the findings indicate that the synergistic use of sustainable and synthetic reinforcements with Al7075 alloys improve both mechanical properties and tribological resistance, making these hybrids is suitable for lightweight, high-performance structural applications.[34]

This research focuses on the mixing of a lightweight high-entropy alloy (LHEA: $Al_{30}Li_{20}Mg_{10}Be_{20}Ti_{10}Cu_{10}$) as a reinforcement phase to improve the mechanical properties of Al7068 aerospace alloy. Composites containing 4, 10, and 16 wt% LHEA have been synthesized and characterized. X-ray diffraction analysis report confirmed the formation of BCC and FCC phases found in the LHEA, along with $CuAl_2$ and $MgBe_{13}$.[35]

Microstructural analysis revealed that particle dispersion strongly dependent on reinforcement content: sparse at 4%, uniform at 10%, and mixed with clustering at 16%. These morphologies, along with phase constituents, directly influenced the mechanical behavior. The composites showing improved tensile strength, plastic deformability, and strain-hardening property. The study concludes that the 10 wt% LHEA composition provided the most effective strength and ductility and to optimized particle distribution and phase evolution. This investigation reflected the importance of LHEA-reinforced Al7068 as a high-performance structural material for aerospace applications.[35]

This study discusses the effect of silicon carbide nanoparticles (SiCnp) on the precipitation behavior of Al7075 alloy. The SiCnp/7075Al composite has been synthesized using powder metallurgy techniques, such as shift-speed ball milling, hot pressing, and hot extrusion, followed by T6 heat treatment. Even at a low reinforcement level of 1 vol%, SiCnp have been found to significantly alter precipitation kinetics, provided finer, uniformly dispersed η' precipitates while simultaneously increasing dislocation density. With increasing addition of SiC nanoparticles to the Al7075 alloy, both the yield strength (YS) and ultimate tensile strength (UTS) of the SiCnp/ Al7075 composites showed an excellent improvement. The maximum values, YS of 559 MPa and a UTS of 609 MPa, have been achieved at 1.0 vol% SiCnp reinforcement, representing improvement of 72 MPa and 53 MPa, respectively, compared to the unreinforced matrix. [36]

Fractographic and compositional analyses revealed that precipitate-induced crack initiation governs the fracture process. The outcomes showed that nano-reinforcement provides a viable pathway to tailor precipitation phenomena. [36]

The importance of advanced materials in enhancing the safety, reliability, and feasibility of space exploration, humanity's is one of the most demanding technological pursuits. Key hazards such as micrometeoroid impacts, fire hazards, space dust, extreme thermal fluctuations, electromagnetic interference (EMI), and the economic burden of space missions are analyzed in terms of material properties. The survey highlights the recent developments in self-healing polymers, flame- and heat-resistant systems, thermal management materials, self-cleaning coatings, EMI shielding solutions, and multifunctional carbon-fiber composites. So, these innovations provided significant outcomes for mitigating risks, extending the operational life of space technologies, and reducing mission costs, and given a roadmap for material-driven advancements in future space exploration.[37]

This review highlights the critical factors affecting the performance and reliability of aerospace bearings which is operating at very high speeds and extreme temperatures. It discusses about material selection, bearing design, manufacturing methods, and surface treatments influence bearing life and performance for industrial applications. This study also explains the role of advanced bearing materials and heat treatment processes, such as nitriding to improve the fatigue life, wear resistance, and tribological behavior under demanding aerospace operating conditions.[38]

According to this author, composite materials provided high strength and stiffness advantages for aircraft structures, but its high life-cycle cost remains a major challenge for large-scale adoption for industrial applications. While many studies have examined costs at individual stages, an integrated understanding is still lacking. In this work, tensile failure tests have been carried out on aircraft-grade reinforced composites, and scanning electron microscopy was used to examine the microstructural features of both transverse and longitudinal sections.[39]

This study examines the microstructural changes that occur when laser cladding is used to repair aluminum alloy 7075 aircraft components. The microscopic analysis showed that laser heating leads to the formation of various precipitates in the cladding layer and causes solute redistribution in the heat-affected zone. The heat-affected portion showed reduced nanoparticle density and segregation of alloying elements at grain boundaries, but with the local loss in strength. while laser cladding is effective treatment for repair, careful control of thermal effects is essential to maintain the mechanical integrity of the repaired material.[40]

This study analyzed the Al6061 reinforced with zirconia micro particles by using an ultrasonic stir casting process. The investigation has been done by Taguchi–Grey response surface method. Important processing parameters such as ultrasonic power and time, stirring temperature and pressure, and reinforcement content were systematically analyzed. Mechanical properties including tensile strength, hardness, elongation, and grain size were used to evaluate performance. Statistical analysis and response surface modeling identified optimal processing conditions that refined grain structure and improved mechanical properties. The results confirm that ultrasonic stir casting process is an effective and reliable technique for producing high-quality aluminum nanocomposites.[41]

This work demonstrates simple and operative fabrication method for producing aluminum matrix composites reinforced with nano-sized α-Al_2O_3 particles. The uniformly distributed alumina nanoparticles must improve strength, hardness, and ductility without introducing porosity or surface defects. The composites show gains growth in tensile strength and elongation as well as compared with pure aluminum, while retaining good workability.[42]

This study focuses on producing Al5059 aluminum alloy metal matrix composite by reinforced with B_4C and Al_2O_3 micro particles with the help of the stir casting method. The results show that mixing of B_4C and Al_2O_3 micro particles significantly improves tensile strength and surface hardness, making the material stronger and more wear resistant. However, as the microparticles content increases, the material becomes less ductile and more brittle, leading to reduced elongation.[43]

This study examined the effects of mixing small amounts of copper on the microstructure and mechanical properties of Al7075 aluminum alloy produced by stir casting. Copper additions between 0.3 and 0.7 wt.% led to noticeable grain refinement and the formation of well-distributed intermetallic phases. As the copper content increased, both hardness and tensile strength improved, with the 0.7 wt.% Cu alloy showing the highest values. Microstructural analysis confirmed clearer grain boundaries and reduced segregation. So copper addition effectively enhances the strength and hardness of Al7075 alloy, making it more suitable for demanding industrial applications.[44]

Researchers compared the mechanical and laser paint stripping methods on aluminum 2024 aircraft structure. Mechanical stripping makes positive impact to remove deep scratches, protective oxide layer completely, and reduced corrosion resistance. laser paint stripping proved to be a safer and more effective method, as it minimized substrate damage while enhancing corrosion protection.[45]

This research has been carried out on aluminum alloy composites reinforced with titanium and carbide based microparticles. This study highlights the common manufacturing methods and the role of reinforcements like titanium and niobium carbide and summarizing work related with mechanical and tribological improvements. It also points out existing limitations and identifies future research directions needed to fully exploit the potential of aluminum-based hybrid MMCs.[46]

Magnesium matrix composite reinforced with fly ash fillers by varying the weight percentages by 2.5%, 5%, and 7.5% and then compared with pure magnesium. At 7.5 wt % fly ash composites was being showed good hardness and wear resistance and its tensile strength has improved by 42%.[47]

This research focuses on how different existing strengthening techniques are applied to the mechanical characteristics of Al7075/TiB_2 and fabricated by ultra sonic treatment and squeeze casting. Analytical methods for the prediction of the elastic modulus and yield strength of metal matrix composite in the presence of porosity were proposed after that analysis have been done on the effects of various reinforcing mechanisms on the composite properties.[48]

This author has been done work on Al6061-T6 to explore the importance of aluminum alloys in aerospace and space applications due to its light weight, high strength, and ease of fabrication. It reviews commonly on aluminum alloy series and explains how different composition, processing, and heat treatment influence performance. Special emphasis is given to Al6061-T6, which offers an excellent balance of strength and durability, making it especially suitable for spacecraft and satellite structural components. Overall, the study helps in selecting and tailoring aluminum alloys to meet specific aerospace mission requirements.[49]

This study shows that combination of carbon nanotubes reinforced with γ-Al_2O_3 nanoparticles can produce strong and highly ductile aluminum composites. The bonding between the reinforcements and the aluminum matrix promotes high dislocation activity during deformation, which delays fracture and allows the material to stretch significantly before failure. The tensile strengthening and toughening effect of CNTs and γ-Al_2O_3 nanoparticles explains the unusually high ductility achieved at high strength levels. The findings showing the strong potential of hybrid reinforcement strategies for developing next-generation, high-performance aluminum metal matrix composites.[50]

Silicon nitrides (Si_3N_4) were used as a reinforcement to improve the mechanical performance of aluminium metal matrix composites. Five samples have been made through stir casting technique. Experimental work has been performed on UTM machine and Brinell hardness testing machine as per ASTM standards to analyzed the ultimate tensile strength and young's modulus by vary the percentage of silicon nitrides from 0 wt% to 10 wt%. The results showed that hardness 44.8 BHN to 54.4 BHN & tensile strength 127.1 MPa to 178.5 MPa has been increased by the addition silicon nitrides and ductility is reduced. Further increase in Si_3N_4 after 8 wt% there is reduction in tensile strength.[51]

By using nanoparticles zirconia (ZrO_2)/ A356.0 aluminium investigation were performed. As usual stir casting method has been used to fabricate the nanocomposites. The results present that hardness improved from 71.33 HBN to 87.63 HBN by increasing the content of nanoparticles from 0% to 5%. Also, improvement in yield strength from 24.52% to 28.9% by varies ZrO_2 from 2.5 % to 5%. Fractured surfaces analysis has been done by SEM.[52]

Studied the effect nanoparticles SiC on Al6061 and composites were fabricated by stir casting process. Nano powder SiC have been added from 3, 5 & 7 wt% and mechanical properties were analyzed.[53]

In this work, Al2024-based hybrid nanocomposites have been used reinforced with B_4C and h-BN nanoparticles were successfully fabricated by using powder metallurgy. The B_4C particles were uniformly dispersed in the aluminium matrix, while h-BN tended to concentrate along particle boundaries. Although increasing h-BN nanoparticles content slightly reduced density and increased porosity, it immensely enhanced wear resistance. It has been observed that composite containing 2% h-BN showed much lower wear loss compared to 1% h-BN, confirming the beneficial role of h-BN in improving tribological performance.[54]

This study investigates Al7075 alloy reinforced with nano-sized Al_2O_3 particles (20–30 nm) fabricated by using an electric stir casting method. Unlike conventional mechanical stirring, electric stirring improved nanoparticle distribution and reduced porosity. The addition of 1 wt.% magnesium enhanced wettability between the matrix and reinforcement. Microstructural analysis using optical microscopy and SEM confirmed that uniform particle dispersion and refined grain structure. As a result, the Al7075/1 wt.% Al_2O_3 nanocomposite exhibited notable improvements in hardness, tensile strength, and impact resistance compared to the base alloy.[55]

This author reviews the previous researchers of composite materials because composite material offered excellent mechanical properties and relatively simple processing methods. The available various fabrication techniques, powder metallurgy plays a key role in producing high-quality composites. In Al–B_4C composites, the reinforcement particles bear the heavy load applied through strong interfacial bonding, enabling the material to withstand higher stresses. Since this transfer load is mainly occurs at the Al–B_4C interface. This paper reviews important studies on Al–B_4C composites fabricated by using powder metallurgy, with a focus on interfacial behavior and its influence on composite mechanical performance.[56]

This study investigated the hardness and wear rate of aluminium matrix Al7075 composites reinforced with Al_2O_3, SiC produced by stir casting. The fabrication of aluminum-based metal matrix composites (MMCs) for its high production efficiency by using the stir casting method. Taguchi L9 orthogonal array have been used to optimize the critical process parameters. The author took Al2O3 (2% constant), SiC (3%, 6% and 9%) composition and heat treatment temperature (140 °C, 160 °C and 180

°C), while optimum wear rate corresponded to 3% SiC and heat treatment temperature 140 °C. 2.2% of ware rate difference was observed.[57]

This review focuses on the use of carbon nanotubes as a reinforcement in aluminium metal matrix composites due to its exceptional mechanical and thermal properties. The performance of Al–CNT composites strongly depend on the amount of CNT added to the aluminium matrix. The paper mainly highlights how different CNT concentrations influence the mechanical properties like tensile strength, yield strength, and compressive strength, showing its potential for lightweight and high-performance aerospace and engineering applications.[58]

This investigation has been done on the development of Al6061-based hybrid nanocomposites reinforced with a constant 5 wt% graphite (Gr) and varying SiC nanoparticle contents (5, 7.5, and 10 wt%, 50 nm), synthesized through a liquid-state stir vacuum die-casting process. The composites have systematically analyzed for physical properties (density and porosity), microstructural, mechanical parameters (tensile, yield, strain, and microhardness), and wear characteristics. From the fabricated specimen, the composite contains 10 wt% SiC + 5 wt% Gr (sample 4) exhibited the excellent mechanical properties, characterized by low porosity (0.72%), enhanced hardness (86 ± 1.1 HV), yield strength of 118 ± 0.5 MPa, and tensile strength of 147 ± 0.58 MPa with a ductility of 3.2 mm strain. Furthermore, this composition has been attained improved tribological behavior, with a reduced wear rate of 0.321 mg/m and a stable coefficient of friction (0.41) under 40 N load at 0.75 m/s. This study validated the efficacy of hybrid nano-reinforcement and advanced casting techniques in tailoring the mechanical and tribological performance of Al6061 composites for various engineering applications.[59]

This study demonstrates that Al6063 aluminium alloy reinforced with B_4C nanoparticles can be effectively produced by using ultrasonic casting process. The nanoparticles have been uniformly dispersed, noticeable improvements have been observed in hardness, yield strength, and tensile strength. However, ductility decreased with increasing content of B_4C. Fracture analysis revealed the failure mechanisms, confirming that nano-reinforcement significantly enhanced the strength while slightly decreasing elongation, making these composites suitable for high-performance engineering applications.[60]

The growing demand for lightweight and energy-efficient materials has increased interest in aluminium metal matrix composites (AMMCs). These composites are mainly fabricated by using solid-state and liquid-state processing methods. Stir casting is the mostly used liquid-state technique due to its simplicity, low cost, and ability to achieve reasonably uniform particle distribution. Powder metallurgy, a common solid-state method, offers better uniformity but at higher cost. Other manufacturing methods such as compo-casting, squeeze casting, friction stir processing, and spray casting are also used in industrial applications.[61]

This author has done analysis with the use of Al6061 metal matrix composites reinforced with fly ash for enhanced protection against high-intensity radiated fields (HIRF). This metal alloy matrix composite, designed with minimal thickness for surface coverage, helps shield aircraft structures and sensitive electronic components. Electromagnetic properties were evaluated in the X-band under normal incidence conditions. The analysis have been done by using MATLAB software, the composite containing 15% fly ash and 85% Al6061 showed a maximum reflection loss of 33.88 dB, indicating strong potential for lightning and electromagnetic interference protection in aircraft applications.[62]

This author presents an effective thermal recycling method to recover carbon and glass fibers from CFRP and GFRP manufacturing waste. Using an optimized process at 550 °C, high recovery rates were achieved while preserving fibre length and quality. The recycled fibres were successfully reused to fabricate new composites, showing noticeable improvements in tensile, stiffness, and impact properties. The results demonstrate that this closed-loop recycling approach is a practical and sustainable solution for reusing composite waste while maintaining the performance of recovered fibres.[63]

Al2024-Ti were used as base metal matrix composite material mixed with SiO_2 nanoparticles (size 30nm). Composites material have been made by stir casting techniques. Micro Vickers hardness testing machine and impact testing machine have been used for examined the strength of the composites. The tensile strength and impact energy significantly increases with the increasing percentage of silicon dioxide.[64]

This author focused on developing Al6351 aluminum matrix composites reinforced with silicon nitride (Si_3N_4) microparticles by using the stir casting process. The silicon nitride (Si_3N_4) microparticles compositions have been used by varying the content of microparticles 0 to3 wt.% and examined to understand the influence on microstructure, mechanical, and wear properties of Al6351. Microstructural analysis report that uniform distribution and proper mixing of Si_3N_4 microparticles within the aluminum matrix. As the Si_3N_4 microparticles content increased, hardness, tensile strength, and compressive strength showed noticeable improvement. Wear testing further show that reduction in wear rate with the addition of higher Si_3N_4. The results demonstrate that adding Si_3N_4 particles effectively enhanced the mechanical strength and wear resistance of AA6351 aluminum composites.[65]

This study focuses on developing new aluminum alloy-based nanocomposite by reinforced with graphene nanoparticles by using the stir casting method by varying the different weight percentages compositions to improve its mechanical and chemical performance. The fabricated composites showed that uniform distribution of graphene, confirmed through density measurements and optical microscopy. Mechanical testing revealed notable improvements in tensile strength and hardness as the graphene content increased, although impact toughness gradually decreased with higher reinforcement levels. The results demonstrate that graphene-reinforced aluminum nanocomposites provide lightweight and high-strength aluminum alloy and is the best alternative to conventional materials like steel and cast iron for advanced engineering applications. [66]

Researcher investigates the thermal and mechanical performance of aluminum-based hybrid metal matrix composites reinforced with graphite and silicon carbide. The composites were produced by using the stir casting method with varying graphite content while keeping SiC constant at 10 wt%. Results showed that adding graphite improved hardness and tensile strength compared to pure aluminum. Additionally, thermal expansion and thermal conductivity decreased with higher graphite content over the temperature range of 50–300 °C.[67]

This author has done analysis on the wear behavior of aluminum alloy composite is strongly influenced by factors such as applied load, reinforcement type, sliding speed, and distance. Understanding how above these parameters is affecting wear mechanisms and helps in selecting appropriate composite systems for specific operating conditions. Overall, the study highlights that proper reinforcement selection and control of service conditions can significantly improve the wear resistance and durability of Al MMCs in dry sliding environments.[68]

This study investigates the mechanical behavior of newly developed CARALL fiber–metal laminate material under both slow and high-speed loading conditions. Tensile tests showed that material strength did not significantly change with strain rate, the failure mechanism differs under dynamic loading compared to quasi-static conditions. High strain-rate testing by using Split Hopkinson Bar highlighted the importance of direct strain measurement through image analysis for accurate results.[69]

Aluminium matrix composites (AMCs) is an advanced improved materials that exhibit excellent structural and mechanical properties compared to monolithic alloys. Its performance is largely determined by the type of reinforcement (continuous or discontinuous) and the fabrication route, which together dictate microstructural integrity, mechanical response, and thermal stability. This review emphasizes processing methods for AMC manufacturing while also outlining its mechanical behavior and its application potential in aerospace, automotive, and defense sectors. Finally, challenges related with commercialization, industrial viewpoints as well as outlines future research directions has been highlighted to guide the progression of AMCs toward broader engineering utilization. [70]

Authors studied layer wise theories and zig-zag models to explore fundamental deep concept of used in modelling.. It discusses key modelling approaches, material models, and numerical techniques used to simulate cracking, delamination, impact, and blast loading. The study highlights LPT and FSDT are not found best for computations of transverse shear of thick and thick laminated plates. Higher order models are capable to forecast the transverse shear effects but need more computational effects. [71]

This study shows that addition of TiO_2 nanoparticles by varying the volume percentage by 0.5, 1 & 1.5 reinforced with Al2024 alloy using a novel powder injection. With increase the volume percentage of TiO_2

nanoparticles, dendrite size decreased and the hardness of samples increased. The Rockwell B hardness was increased by 31.57% at 1.5 volume percentage of TiO_2 nanoparticles. The elongation of samples was reduced with increasing the nanoparticles content. The sample with 1.5 volume percentage of TiO_2 showing 13.8% reduction in elongation compared to the unreinforced sample. The influence of TiO_2 content on the mechanical properties, porosity, microstructure and wear resistance of the samples have been investigated. SEM images showed that TiO_2 nanoparticles uniformly distributed into the metal matrix The nanoparticles were uniformly dispersed in the matrix, leading to finer dendritic structures and a clear increase in hardness and wear resistance. While strength improved notably up to 1% TiO_2 addition, higher content caused a slight reduction in ductility due to increased particle presence.[72]

Taguchi method has been used for set the design parameters such as Al_2O_3 (2%), SiC (3%, 6% & 9%) and heat treatment temperature (140 °C, 160°C and 180 °C) to found the hardness and wear rate of Al7075/ Al_2O_3/ SiC metal matrix composite. The samples were fabricated by stir casting techniques. Vicker's micro hardness testing machine and pin-on-disc wear testing machine were investigated the strengths. The results have been showed that SiC at 9%, Heat treatment temperature at180 °C is optimum values for hardness and SiC at 3%, Heat treatment temperature at 140 °C is an optimum value for wear rate.[73]

Particulate-reinforced aluminum metal matrix composites (Al-MMCs) have gained significant attention due to its low density and the capacity to achieve tailored performance combinations, including high specific stiffness, strength, and improved creep resistance. Recent advancements over the past period highlight the role of micro and nano particle size, morphology, and distribution, as well as the grain structure of the aluminum alloy matrix, in governing the mechanical response of these systems. Particular focus has been done on MMCs fabricated using heat-treatable aluminum alloys. From these developments, B_4C-reinforced 7xxx series (Al–Zn–Mg) alloys have shown capacity, representing improved strength and durability, while also revealing processing and microstructural constraints that allow future research work for industrial deployment.[74]

This work focuses on the fabrication and tribological testing of aluminum–fly ash metal matrix composites produced by using the stir casting technique. Fly ash has been added to aluminum in varying amounts of 2%, 4%, and 6% by weight. Researcher was found that the MMC with 6% weight of fly ash content gives results low wear (0.32 g) and 4% weight of fly ash content gives the low coefficient of friction (0.12). These results indicate that fly ash is an effective and economical reinforcement for improving the wear resistance and frictional behavior of aluminum composites.[75]

In this research work, the mixing of ceramic nanoparticles such as Al_2O_3 into 7xxx-series alloys and it can refine the microstructures and enhanced the mechanical strength. This study introduced an innovative fabrication route in which Al_2O_3 nanoparticles has been added into molten Al7075 alloy via argon gas flow, followed by solid–liquid mixed casting. This fabrication method ensured homogeneous particle dispersion and reduced clustering. Tensile strength and hardness have been significant improved as compared to the unreinforced alloy, with optimal performance attained at 1.5 wt.% Al_2O_3. Beyond this composition content, properties degradation was observed, attributed to nanoparticle agglomeration and related defects. This outcomes showing both the benefits and critical processing route of nanoparticle reinforcement providing valuable inside information for tailoring high-performance Al-based metal matrix composites.[76]

The new casting set up ball milling, and semi-solid stirring method were used to fabricate the Al7075 composites reinforced by nanoparticles Al_2O_3. The results showed that tensile strength and hardness of Al7075 were increased with addition of Al_2O_3 from 0% to 1.5% and fine the microstructure. [76]

Aluminum alloys, particularly Al7075, are widely used in the application of aerospace and automotive applications due to its low density, formability, and corrosion resistance; however, further strengthening is mostly required for advanced structural demands. This study examined the Al7075 reinforced with 2.5 wt.% and 5 wt.% TiC particles to enhanced its microstructural and mechanical performance. By addition of TiC micro particles transformed the alloy's dendritic morphology into the more refined non-dendritic structure while simultaneously showing the precipitation of secondary phases such as Al_2CuMg, Al_2Cu, and Fe-rich intermetallics. These microstructural changes have been strengthening via grain boundary and particulate mechanisms, resulting in improved hardness and excellent wear resistance, with wear behavior moving from adhesive to abrasive modes. However, the composites exhibited higher corrosion susceptibility due to precipitate-induced pitting, underscoring the trade-off between mechanical enhancement and electrochemical stability.[77]

In this study, aluminium–graphite nanoplatelet (xGnP) composites has been fabricated by using the powder metallurgy method. The addition of xGnP improved the wear resistance and the excellent mechanical properties has been observed at 3 wt% graphite nanoplatelets reinforcement. Hardness increased up to 1 wt% xGnP but decreased at higher contents, while tensile strength gradually reduced due to the formation of interfacial Al_4C_3 particles. Overall, the outcomes show that controlled addition of xGnP can be effectively enhance wear performance, though excessive reinforcement negatively affects mechanical strength.[78]

High energy milling and spark plasma sintering methods have been introduced to made Al2024-TiN nanocomposites. Researchers have been studied the impact of TiN on the microstructure of Al2024. Its grain sizes were refined from 87 nm to 30e50 nm when TiN particles added. Al2024-2TiN nanocomposite show excellent compressive properties (yield strength = 730 MPa and ultimate strength =871 MPa) which are 185% and 35% more than without TiN.[79]

Nanocomposite Al/Al_2Cu were fabricated through friction stir processing (FSP). Nanocomposite were investigated by FPS parameters like travel velocity, rotational velocity, number of FSP passes, and pin profile of the microstructure, chemical reaction and hardness of Al based nanocomposite. Pin diameter has a more impact on its microstructure and hardness values. The hardness values have been improved by 57% compared with the base metal.[80]

Studied the effect SiC nanoparticle on the Al356. Authors have made the nanocomposite material by stir casting process and then varied the SiC from 1% to 5% by weight. With the addition of nanoparticle have been enhanced the mechanical properties of A356 like Hardness (increased by

125%), tensile strength (improved by 41% in YTS and 45% in UTS) and compressive strength (improved by 603 MPA). Toughness, elongation and forgeability have been decreased by the addition of SiC nanocomposites. The fractured surface of A356/SiC nanocomposites have been studied by SEM and EDX spectrum.[81]

Aluminium matrix hybrid composites have been used with microparticles SiC (10 wt %) and nanoparticles TiB_2 (taken content 1 wt %, 3 wt % & 5 wt %). The AMHCs were fabricated by metallurgy process. SEM and TEM have been used for examined the microstructure and mechanical performance of composites. The analysis indicates that strength of the hybrid composites was being increased when the concentration of TiB_2 is increased.[82]

Al6061 metal matrix composites reinforced with nanoparticles B4Cp have been fabricated by ultra-sonic cavitation casting techniques. With the addition of boron carbide in Al metal matrix composite were increased the hardness, tensile strength also maintains the ductility and impact resistance of Al alloy. It's also refined the matrix grains of microstructure of metal matrix composite. The results obtained by optical microscopy.[83]

Hardness and sliding test have been adopted to study the mechanical behavior of aluminum matrix composites by using two reinforcement TiB_2 and TiO_2 nanoparticles. Comparison has been performed between two nano powders in terms of distribution of nanoparticles. Higher strength showed by 356 Al- TiB_2 as compared with 356 Al-TiO_2[84]

Authors have improved the mechanical properties of Al/CFRP laminates and fabricated by carbon fiber reinforced plastic (CFRP) with aluminum sheet. Explored the effects of various toughening methods such as acid etching and mechanical pattering for surface treatment of Al alloy sheet, and nanofiller addition for interfacial enhancement. Double cantilever beam tests proved that the improvement on the interlaminar mechanical properties of Mode-I fracture in terms of higher critical load P_C and fracture toughness G_{IC}. Experimental and numerical study confirmed that improved bonding and crack resistance at the interface are the main reasons for the enhanced mechanical performance.[85]

Researcher studied the development of aluminum matrix composites by using a cost-effective stir casting process with hybrid reinforcement of fly ash and alumina. The use of industrial waste fly ash with aluminum alloy not only reduces material cost but also decreases the density of the composite. Mechanical tests such as tensile, compressive, and hardness evaluations shown that combining fly ash with alumina significantly improves the mechanical performance of the composites. The results show that hybrid composites exhibit better strength and hardness at lower densities, making it suitable for lightweight and advanced engineering applications.[86]

This study focuses on improving the wear and tribological behaviour of Al-Si10Mg aluminum alloy composites. An Al-Si10Mg alloy reinforced with alumina and graphite were fabricated by using liquid metallurgy and tested under dry sliding conditions. With the help of Taguchi method and pin-on-disc wear testing, the effects of load, sliding speed, and reinforcement content has been analyzed by selected L27 Orthogonal array. Results showed that hardness increased with higher graphite content, while wear rate and friction were strongly influenced by applied load. Overall, the hybrid composite demonstrated improved wear resistance under optimized conditions, making it suitable for applications requiring better durability and friction control.[87]

In this paper, mechanical and wear properties of the hybrid metal matrix composites have been studied. By stir casting method Mica and SiC ceramic particles have been mixed into Al356. Effects of its ceramic particles on Al356 were examined by SEM and EDX Spectrum. Microstructures have been studied by SEM and chemical composition has been studied by EDX. Its results shown that improvement in strength and hardness were being attained by Al/10SiC- 3mica composites[88]

The addition of ZrB_2 particles on the microstructure and mechanical properties of Al6061 alloy was studied in this work. Researcher have been developed Al6061 alloy particulate reinforced metal matrix composites (PMMCs) are prepared by adding zirconium boride (ZrB_2) particles by the in-situ reaction of K_2ZrF_6 and KBF_4 salts with molten aluminium. ZrB_2 particles significantly refined the microstructure and enhanced the mechanical properties of AA6061 alloy. The varying percentage composition of ZrB_2 from 0 to 10 in steps of 2.5. The results shown that hardness, ultimate tensile strength and wear resistance of Al6061 alloy was noticed with the rise in ZrB_2 content [89]

This study examines the dry sliding wear behavior of aluminium alloy composites reinforced with silicon carbide, with and without solid lubricants such as graphite and antimony trisulfide (Sb_2S_3). The composites were fabricated using hot powder preform forging and its wear performance was evaluated under different loads and sliding speeds. Results show that the hybrid composite containing solid lubricants exhibits lower wear rate and better tribological performance than the binary SiC-reinforced composite. When compared with iron-based MMCs, the aluminium hybrid composite shows slightly higher wear but offers quieter operation and overall acceptable tribological behavior, making its lightweight material alternative.[90]

This paper compares latest manufacturing methods for AW-AlCu4Mg1(A) aluminium matrix composites reinforced with Al_2O_3 particles. The composites were produced using two techniques: powder metallurgy and pressure infiltration of porous ceramic preforms. The results indicate strong indication for achieving desirable mechanical and structural properties. These composites are particularly excellent for automotive applications, although further research is needed before large-scale use. The developed manufacturing techniques are especially suitable for producing near net-shape components and locally reinforced parts.[91]

This study focused on aluminium matrix composites with a high content of silicon carbide particles produced using a pressure less infiltration method. The composites were evaluated in fabricated condition for microstructure, phase composition, density, porosity, and thermal expansion behavior. As the SiC content increased, density showed a linear increase, while the coefficient of thermal expansion decreased nonlinearly. These trends highlight how a high volume of ceramic reinforcement effectively improves dimensional stability, and the experimental results showed good agreement with established composite property models.[92]

Researchers studied the fracture toughness and fracture behavior of COMRAL-85™, 6061 aluminium alloy reinforced with 20 vol% Al_2O_3 alumina ceramic microspheres and produced by using liquid metallurgy. Fracture toughness tests using short rod and short bar (chevron-notch) specimens machined from extruded 19 mm diameter rod, showed that toughness decreased with material particle volume fraction ranging between 5% and 30%. Additional tests on powder-metallurgy composites revealed that fracture toughness decreased as the ceramic particle content increased, though the rate of reduction slowed at higher volumes.[93]

Researcher has studied the corrosion behavior of an Al_2O_3 fibres – reinforced magnesium (AS41) metal matrix composite in salt solutions of varying NaCl concentration and pH value. The following test have been conducted such as electrochemical tests, hydrogen evolution measurements, and microscopic analyses and the composite's corrosion behavior were compared with that of the pure AS41 unreinforced magnesium alloy. The outcomes show that the composite exhibited corrosion behavior similar to the base alloy, although its corrosion resistance decreases with the increasing contents of chloride. The role of alumina fibers in influencing corrosion mechanisms is clearly explained, and a corrosion model is proposed to describe the observed behavior.[94]

In this study, aluminium metal matrix composites reinforced with silicon carbide and graphite were fabricated by using the liquid metallurgy process. The dry sliding wear behavior of the hybrid composite has been investigated and compared with that of Al/SiC composite. Taguchi's experimental design and ANOVA has been used to study the effects of load, sliding speed, and sliding distance on wear performance. The results show that graphite microparticles acts as an effective solid lubricant, significantly improving the dry sliding wear resistance of Al/SiC composites.[95]

This paper investigates the effect behavior of aluminium and SiC particle–reinforced aluminium matrix composites under different temperature range. Charpy impact tests has been conducted on extruded and heat-treated samples made from aluminium alloys 2124, 5083, and 6063, reinforced with different SiC particle sizes and extrusion ratios specimen at temperatures varying from 176 °C to 300 °C. The results show that impact performance is strongly influenced by particle distribution, particle cracking, and the quality of bonding between the matrix and reinforcement. Particle agglomeration reduced impact strength in Al2124 and 6063 composites, while Al 6063–based composites exhibited the best overall impact resistance. Larger SiC particles and higher extrusion ratios improved the impact strength of 6063 composites. Due to temperature variation had only low effect on impact behavior, indicating stable performance across the tested temperature range.[96]

Researcher has studied fiber-metal laminates (FMLs) made from the different combinations of glass fiber/epoxy layers and aluminum or steel sheets to investigated FMLs capability and used as alternative for conventional aerospace materials. The results show that FMLs provided balanced combination of low weight, high strength, and excellent fatigue and impact resistance, making it strong suitable for aircraft structures when compared to traditional metals and fiber-reinforced composites.[97]

Researcher developed finite element model to simulate the orthogonal machining of alumina-reinforced Al6061 metal matrix composites with help of tungsten carbide tool. This FEM model took temperature-dependent material properties and particle–matrix interface failure. It successfully predicts cutting forces, stress distribution on alumina particles, interface damage, and cutting temperatures, showing good agreement with experimental results. This outcomes provided valuable information about particle failure and stress behavior during machining of 6061 aluminium MMCs.[98]

Researcher studied about friction stir processing (FSP) affects the mechanical and fatigue behavior of 2618 aluminium alloy (Sheets of 7 mm thickness) reinforced with 20% alumina particles. After T6 heat treatment, the processed sheets showed significant grain refinement in the nugget zone, leading to improved tensile, fatigue, and superplastic properties. High-temperature tensile tests have been performed to study the high temperature behaviour of the 2618 aluminum alloys in longitudinal direction. Fatigue test is conducted to analyzed fracture surfaces by SEM.[99]

This study investigates the corrosion behavior of four aluminium metal matrix composites (A360/SiC/10p, A360/SiC/20p, A380/SiC/10p, A380/SiC/20p) reinforced with SiC particle contents deep in 1–3.5 wt% NaCl at 22 °C. Scanning electron microscopy (SEM) and X-ray diffraction (XRD) has been used to analyzed the corrosion behavior. The outcomes shows that matrix alloy composition plays a more significant role in corrosion resistance than the amount of SiC reinforcement or salt concentration.[100]

Aeroengines are being operated under extremely harsh conditions, demanding materials that type of material to withstand high temperatures, stresses, and fatigue while remaining lightweight and cost-effective. To meet following requirements for higher thrust and efficiency, advanced materials such as metal matrix composites (MMCs) are being explored. This chapter highlights the potential advantages of aluminium and titanium matrix composites in aeroengine applications, discusses about its manufacturing methods, and addresses the challenges involved.[101]

Researcher examined the dry sliding wear behavior of ZA-27 alloy composites reinforced with garnet particles produced through liquid metallurgy. The results show that adding garnet (2 to 6% in steps of 2 wt.%) improved the wear resistance, with higher garnet content lowered the wear rates compared to the base alloy. It is found that wear rate increased with higher load and sliding speed for all samples, reinforced composites performed throughout better. At higher loads, the wear mechanism shifted from abrasion to particle-induced delamination. The surface structure analysis have been explained using scanning electron microscopy (SEM) analysis of the worn surface.[102]

Researcher investigated about the manufacturing method and mechanical properties of 6061 aluminium alloy composites reinforced with both coated and uncoated silicon carbide particles. Mixing of SiC particles with copper considerably improved the interfacial bonding properties, leading to higher failure strains and better overall mechanical performance. Clear differences in fracture behavior were observed between coated and uncoated reinforcements, highlighting the role of interface quality. SEM tensile testing showed that enhanced failure strains in samples containing copper-coated silicon carbide reinforcements compared with their non-coated. The results also show that particle size plays important role in determining the mechanical properties of the composites.[103]

Researchers explore the future of the advanced materials and its scope for industrial application. This paper describes the key features such as microstructural and property characteristics of aluminum metal matrix composites of various grades of aluminum alloys and suggesting various applications. [104]

This study examines the possibility of recycling a silicon carbide reinforced aluminium composite using an innovative disintegrated melt deposition technique. Microstructural analysis showed only minor reductions in porosity and reinforcement size, with no change in particle distribution and improved bonding at the matrix reinforcement interface after recycling. The aluminum alloy reinforced with SiC exhibited higher interfacial hardness, improved strength, stiffness, and ductility, along with a lower coefficient of thermal expansion compared to the original material. The outcomes show that physical and mechanical property characterization revealed an increase in elastic modulus, 0·2% yield strength, ultimate tensile strength, and ductility, and coefficient of thermal expansion of the recycled specimens is reduced when compared with the parent composite.[105]

Researchers discussed about the surface treatments and coatings on reinforcements such as graphite, silicon carbide (SiC), and alumina (Al_2O_3) may have improved interfacial bonding and overall composite properties. Metallic coatings enhance wettability but may alter the matrix composition through alloying, while ceramic coatings act as diffusion barriers that limit harmful interfacial reactions. Multilayer coatings offer multiple benefits, including improved wetting, stress relief, and interfacial stability.[106]

This paper investigated the sliding wear behavior of titanium carbide (TiC) reinforced aluminium metal matrix composites produced by using novel casting technique that enabled strong particle–matrix bonding. The wear rate of these composites has been found under different loads and particle volume fractions and compared with a conventional SiC-reinforced composite. These results show that TiC reinforcement significantly reduces wear rate and delays the transition to severe wear. From the following samples, TiC-reinforced A356 alloy has gained the best wear resistance due to its higher hardness and grain refinement effect. While increasing particle content reduced composite wear, it also increased wear of the steel counterface, indicating that an optimal particle volume fraction is required to achieve balanced and minimal overall wear.[107]

2.3 Literature Review on Al7068

Al7068 nanocomposites reinforced with Si_3N_4 (0.5%, 1%, 1.5% & 2 wt.%) were fabricated using ultrasonic-assisted stir casting. Pre-processing through planetary ball milling ensured uniform particle distribution. Microstructural analysis shows that grain refinement with increasing reinforcement, though agglomeration occurred at higher content. Optimal mechanical properties were achieved at 1.5 wt.% Si_3N_4, showing significant improvements in hardness, tensile, and compressive strength. [108]

This research investigates the development of hybrid aluminium matrix composites (HAMCs) using the stir casting technique. Aluminium alloy Al7068 was reinforced with varying weight percentages of silicon carbide (SiC: 0–8 wt%), along with a constant 4 wt% coconut shell fly ash (CSFA) and 2 wt% nano-magnesium. Hardness improved significantly, with an increase of about 32% when 8% SiC was added along with CSFA. Tensile strength increased by nearly 65% at 6% SiC + 4% CSFA reinforcement. The addition of reinforcements (SiC + CSFA) generally led to a reduction in ductility as well as impact energy reached up to 3.2 J in composites containing both CSFA and SiC.[109]

These phase constituents and particle distribution morphologies of this research improved the mechanical properties, including tensile strength, plastic deformability, and strain hardening, while simultaneously reducing the necking susceptibility. The strengthening mechanisms and micromechanical report discussed in this study highlights the part of phase changes and particle distribution in governing the composite's behavior. The findings show that 10 wt% LHEA reinforcement gives the optimal value of strength and ductility, making it as the most effective composition for aerospace-grade Al-7068 hybrid composites. [24]

Zirconium dioxide (ZrO_2) proves to be an effective reinforcement due to its strong interfacial bonding with the Al7068 matrix. Microstructural examination confirms a uniform distribution of ZrO_2 particles along with good matrix–reinforcement adhesion. In this research, Al7068 MMCs has been synthesized via stir casting with varying ZrO_2 contents (0–8 wt%), followed by microstructural, fractographic, and mechanical evaluation. As the composition of ZrO_2 increases, the tensile strength and hardness of the metal matrix composites have been improved, indicating that the micro particles significantly strengthen the aluminum alloy. However, the percentage elongation decreases with increase of ZrO_2 content, showing increased brittleness caused by the presence of hard ceramic phases that showing localized crack initiation. Overall, the composite containing 8 wt% ZrO_2 exhibits a 13% improvement in tensile strength and a maximum hardness increase of about 40% compared to the unreinforced Al7068 alloy.[110]

Researcher studied the effect of Equal Channel Angular Pressing (ECAP) on the mechanical properties of Al7068 aluminium alloy. The alloy samples have been processed through ECAP split die with a 90° channel angle for up to four passes using route BC, where the specimen is rotated in the same direction after each pass. This process was carried out at an optimized temperature of 200°C. Both unprocessed and ECAP-processed samples have been tested for tensile strength and Vickers microhardness. The results showed that ECAP-processed samples showed higher tensile strength and hardness compared to the unprocessed samples, indicating a clear improvement in mechanical performance.[111]

The tribological behavior (wear performance) and impact strength of Al7068 aluminium alloy has been investigated by varying the percentages of two alloying elements such as Magnesium (Mg) and Zinc (Zn). The four different samples of Al7068 aluminum alloy have been tested by varying the Mg from 2.2% to 3% and Zn from 7.3% to 8.3%. Wear rate test results showed that the alloy with 3% Mg and 7.6% Zn gained the lowest wear rate under loads of 2 kg and 3 kg, indicating better wear resistance. In contrast, the composition with 2.75% Mg and 7.3% Zn showed the highest wear rate at a lower load of 1 kg. Impact test results given that the composition with 2.2% Mg and 7.6% Zn demonstrated the highest impact strength, suggesting improved toughness and overall performance of the alloy.[112]

Researcher focuses on the fabrication of Al7068 composites by using the powder metallurgy route with rice husk ash (RHA) (0%, 4%, 6% & 8%) and silicon carbide (SiC) (0%, 4% & 6%) as reinforcements in different weight percentages. The composite powders have been compacted under a pressure of 400 MPa and sintered at 720°C for three hours.

Microstructural analysis was carried out using SEM and EDX techniques. Hardness and compression tests have been performed to evaluate mechanical performance. The results showed that hardness improved with an increase in SiC content, but decreased with the addition of rice husk ash.[113]

Table 2.1 Summary of various reinforcements with aluminium alloy metal matrices.

S. No	Author's Name /Aluminum Alloy	Fabrication methods	Investigation Approach	Reinforcement	Key Findings	Ref. No.
1	R. Shivashankar *et al (2025)/* Al6061 alloy	stir-casting	deep learning–based DLSTM model	graphite and tungsten carbide	4% Gr and 6% tungsten carbide, which showed notable improvements over the base alloy	[8]
2	S. K. Pradhan *et al. (2025) /* pure aluminium-based 99.9% purity	Ball milling	field emission performance	graphene	2 wt% graphene exhibits the low field emission performance, whereas other compositions, including 1.25, 1.5, and 1.75 wt%, show comparatively higher values	[10]
3	A. Shanmugam *et al. (2024)/* Al7075	Al7075	abrasive water jet machining	Taguchi and DEAR methods	optimal machining condition was identified as 180 MPa pressure, 3 mm standoff distance, and 40 mm/min traverse speed, which resulted in noticeable	[16]

S. No	Author's Name /Aluminum Alloy	Fabrication methods	Investigation Approach	Reinforcement	Key Findings	Ref. No.
					improvements in surface quality	
4	K. R. V. Gudipalli *et al. (2024)* / Al7075	Stir casting	Taguchi method, ANOVA	SiC, Graphite	Addition of 9% graphite content reduces both the coefficient of friction and wear rate, Improved self-lubricating properties 18	[17]
5	S. Ahmad *et al. (2024)* / Al-7075	stir casting	pin-on-disc setup	TiO_2	Maximum and minimum modul us values are reported as 98.9 GPa at 25 °C and 77.3 GPa at 350 °C for 8% TiO_2 mix MMC. Higher TiO_2 compositio ns (6–8%) displayed stable tribological performance, while 10% TiO_2 showed increased COF at higher sliding speeds. Higher compositions (8–10%) showed increasing wear rates at higher sliding speeds	[32]
6	J. Sharma *et*	Stir Csting	Tensile	zirconium	micron-sized (1–	[26]

S. No	Author's Name /Aluminum Alloy	Fabrication methods	Investigation Approach	Reinforcement	Key Findings	Ref. No.
	al (2024) / Al7075		Testing, Optical Microscopic	dioxide (ZrO_2)	3 wt%) and nano-sized (0.17–0.51 wt%) ZrO_2 particles. Mechanical testing evaluated peak stress, load capacity, and yield strength	
7	B. S. Lovevanshi *et al.* *(2024) /* Al 7075	Ultrasonic transducer, two-stage stir casting	SEM, Optical microscopy	Synthetic	Compressive strength and micro-hardness were both increased, and the highest values were found to be 569.172 MPa and 178.86 HV.	[34]
8	M. Alipour *et al. (2023)* / Al–10Zn–3.5Mg–2.5Cu	ball milling and stir casting	microstructural changes and mechanical performance	graphene nanoplate (GNP)	GNP addition improved the strength and grain refinement. GNP addition improved the grain refinement	[18]
9	P. Samal *et al. (2023)* / Cu matrix (C87600) -	Stir casting	Microstructural and Fracture analysis , FESEM, XRD	SiC–graphite and WC–graphite	WC–graphite reinforced composites exhibited superior mechanical properties compared to SiC–graphite reinforced ones. Fracture	[23]

S. No	Author's Name /Aluminum Alloy	Fabrication methods	Investigation Approach	Reinforcement	Key Findings	Ref. No.
					analysis indicates that both composites firstly fail through ductile fracture test, confirming its ability to retain toughness while gaining strength	
10	A. A. Akinwande *et al (2023)* / Al6061-T6 alloy	Stir Csting	Using response surface methodology	palm kernel shell ash (PKA), rice husk ash (RHA), and waste steel particles (STP)	2–4% RHA and PKA enhanced tensile and flexural strengths, while 4–6% led to a decline in strength. Meanwhile, 2–6% of the particles enhanced the tensile and compressive strengths and moduli	[24]
11	A. Abebe Emiru *et al. (2023)* / Al6061	Stir Csting	physical, mechanical, tribological behaviour and morphological behaviour	MoS_2, SiC and B_4C	Highest density was observed at Al6061/12% SiC/4% MoS_2 Maximum hardness 114.03 HV showed at 12% B_4C/ 4% MoS_2	[28]
12	S. Kolappan *et al (2022)*	Stir casting	UTM	SiC, Cocunut shell fly ash	Addition of SiC and coconut shell fly ash to	[109]

S. No	Author's Name /Aluminum Alloy	Fabrication methods	Investigation Approach	Reinforcement	Key Findings	Ref. No.
	Al7068				Al 7068 increases the tensile strength of the composites with 6 % and hardness reaches 173 BHN at 8 wt% SiC and 4 wt% CSFA	
13	B. Pu *et al.* *(2020)* / Al7075	shift-speed ball milling, hot pressing, and hot extrusion	SEM, EDX, XRD	SiCnp	YS of 559 MPa and a UTS of 609 MPa, have been achieved at 1.0 vol% SiCnp reinforcement, representing improvement of 72 MPa and 53 MPa, respectively, compared to the unreinforced matrix.	[36]
14	A. Mahadikar *et al.* *(2020)* /Al7068	Stir Csting	tribological behavior and impact strength	Magnesium (Mg) and Zinc (Zn)	Mg from 2.2% to 3% and Zn from 7.3% to 8.3%. Wear rate test results showed that the alloy with 3% Mg and 7.6% Zn gained the lowest wear rate under loads of	[112]

S. No	Author's Name /Aluminum Alloy	Fabrication methods	Investigation Approach	Reinforcement	Key Findings	Ref. No.
					2 kg and 3 kg, indicating better wear resistance. .2% Mg and 7.6% Zn demonstrated the highest impact strength	
15	M. Ravikumar *et al. (2018)* /Al7075	Stir casting	Taguchi method	SiC, Al_2O_3	Optimum value of Hardness SiC at 9% and Wear Rate SiC at 3% , Al_2O_3 keep constant at 2%	[57]
16	M. Madhusudhan *et al. (2017)* Al7068	Stir casting	Vickers hardness test, UTM	ZrO_2	13% increase in tensile strength of AA7068 and 40% hardness improved at 8wt% ZrO2 when compared with Al7068.	[110]
17	W. Ruirui *et al. (2017* Al7075	Ball milling, and semi-solid stirring	SEM, EDX	Al_2O_3	Tensile strength and hardness increase from 0 to 1.5% and then begin to decline when it exceeds 1.5%.	[76]
18	Asiva Noor Rachmayani *et al. (2015)/* Al7068	Stir Csting	XRD	LHEA: $Al_{30}Li_{20}Mg_{10}$ $Be_{20}Ti_{10}Cu_{10}$	Composites containing 4, 10, and 16 wt% LHEA have been	[35]

S. No	Author's Name /Aluminum Alloy	Fabrication methods	Investigation Approach	Reinforcement	Key Findings	Ref. No.
					synthesized and characterized. Formation of BCC and FCC phases found in the LHEA, along with $CuAl_2$ and $MgBe_{13}$	

2.4 Outcomes of Literature Review

A comprehensive review of the literature reveals that most existing studies have focused on the individual reinforcement of aluminum alloys using either Silicon Carbide (SiC) or Silicon Nitride (Si_3N_4). So, several researchers have investigated the effects of Silicon Carbide (SiC) and Silicon Nitride (Si_3N_4) individually on mechanical behavior. The stir casting method is an ideal choice for fabricating metal matrix composites (MMCs) such as the micro SiC/Si_3N_4 particles reinforced Al7068 composite due to its simplicity, cost-effectiveness, and flexibility

Previous studies have been explored many fabrication methods, such as stir casting, powder metallurgy, and squeeze casting, to embed SiC, ZrO_2, Al_2O_3, TiC and Si_3N_4 particles into the Al7075, Al7061 Al7068, matrix.

Research findings propose that different reinforcement like SiC, ZrO_2, Al_2O_3, TiC enhances mechanical properties such as hardness, compressive strength, and wear resistance due to its exceptional stiffness and ability to restrict dislocation movement. On the other hand, Si_3N_4, BN, MgO, graphene reinforcement enhances fracture toughness, thermal stability, and resistance to creep, due to its high strength-to-weight ratio and outstanding resilience to thermal shock.

Specific issues to address include:

1. **Mechanical Property Enhancement:** Al7068 aluminum alloys showing high strength, its performance may be improved further by the addition of silicon carbide and silicon nitrite microparticles. The challenge is being raised to determining the proper compositions of silicon carbide and silicon nitrite content and its distribution that maximizes the composite's strength, hardness, and fatigue resistance without losing its essential properties.
2. **Corrosion Resistance:** The reinforcement of silicon carbide and silicon nitrite may be improved the corrosion resistance of the composite, but the exact mechanisms and effectiveness of silicon carbide and silicon nitrite is need to determine through the investigation.
3. **Manufacturing and Processing:** The addition of silicon carbide and silicon nitrite into Al7068 become challenge for the researchers in manufacturing and processing. It is important to develop effective methods for the uniform diffusion of silicon carbide and silicon nitrite within the aluminum matrix and to evaluate the impact of processing techniques on the final properties of the composite.
4. **Cost and Feasibility:** While silicon carbide and silicon nitrite providing several advantages, its addition must be balanced with considerations of cost and feasibility. It is important to assess the cost implications of using silicon carbide and silicon nitrite as a reinforcement and identify methods to minimize any increment in production costs.

2.5 Motivation

a) **Enhanced Mechanical Properties**: Combining Al7068 with nano and microparticles to form new composites material to improves its mechanical performance, such as fatigue resistance, impact strength, and tensile strength. Engineers can develop these components that can meet the demand of modern industrial and engineering applications by investigations.

b) **Corrosion Resistance**: Although Al7068 is susceptible to corrosion, its performance can be improved when used in composite form. Research on such composites can help to develop such materials with better resistance to environmental conditions as well as increase the service life of components and reduce maintenance costs.

c) **Cost Effectiveness:** Developing Al7068–based composites offer the opportunity to produce high-performance materials at comparatively lower cost. This makes advanced materials more accessible for industrial applications while still maintaining strength and reliability.

d) **Innovation and Development:** Research on Al7068 composites can generate innovation in material science and manufacturing processes. Such studies may lead to the development of new fabrication techniques, improved production methods, and technological breakthroughs that benefit multiple engineering fields.

e) **Eco-Friendly Approach**: Research can also focus on making Al7068 composites more environmentally friendly. This includes exploring recycling methods, reusability, and sustainable production techniques, which can help reduce waste and promote greener industrial practices.

2.6 Strength of Study

The study of Al7068 reinforced with silicon carbide and silicon nitride microparticles is goal to improving the both mechanical properties and corrosion resistance, which are major challenges for the manufacturing of aluminum alloy Al7068. This alloy is well known for its high strength and lightweight characteristics and making it suitable for aerospace and other high-performance applications. However, its poor corrosion resistance and susceptibility to fatigue under repeated loading limit its broader use. To overcome these limitations, silicon carbide and silicon nitride materials recognized for its exceptional hardness, strength, and stability are mixing into the Al7068 matrix. This research investigates that how different proportions of silicon carbide and silicon nitride influence the key properties such as tensile strength, hardness, fatigue resistance, and corrosion behavior of the composite.

The results of this study may be highly beneficial for industries that require lightweight yet strong and corrosion-resistant materials, positioning Al7068/silicon carbide and silicon nitride composites as a promising material for critical engineering applications.

2.7 Objectives of Research

f) **To develop a new high strength** composite material Al7068 Alloy infused with micro silicon carbide and silicon nitride.

g) **To investigate the mechanical properties—**including tensile strength, yield strength, elongation, hardness and toughness of Al7068 reinforced with SiC and Si_3N_4 particles.

h) **To examine the influence of micro-sized SiC and Si_3N_4 reinforcements** on the mechanical properties of the developed composites, both with and without T6 heat treatment.

i) **To investigate the microstructural characteristics** of the developed material using optical microscopy and Field Emission Scanning Electron Microscopy (FESEM).

j) **To analyze the fractographic** of the fractured surfaces using Field Emission Scanning Electron Microscopy (FESEM).

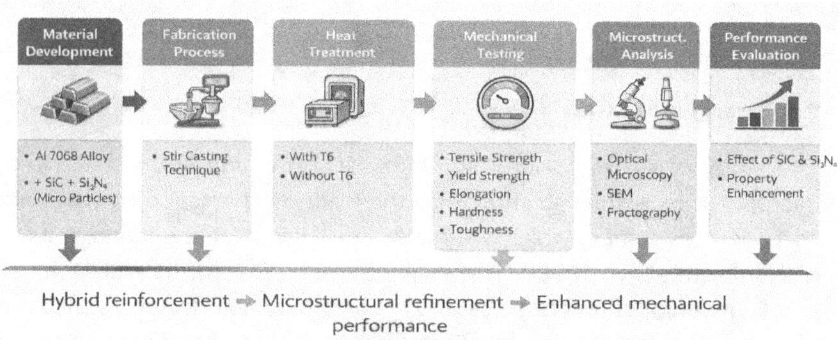

Figure 2. Objective Flow Chart

Chapter - 3
Methodology and Characterization

Chapter 3
Methodology and Characterization

3.1 Materials and Processing Route

Aluminum alloy 7068, recognized for its superior strength-to-weight ratio, has emerged as a prime alloy for high-performance aerospace and defense applications. When use as a matrix material, Al7068 provide an excellent platform for reinforcement with agents such as carbon fibers, glass fibers, boron, or metallic particulates. Al7068 provides a robust and lightweight structure by the mixing of reinforcing agents such as ceramic particles (e.g., SiC, Si_3N_4, Al_2O_3), metallic reinforcements, or hybrid combinations aimed at further enhancing its mechanical performance.

The fabrication of aluminum alloy is commonly carried out through techniques such as stir casting, powder metallurgy, and advanced solid-state processing routes. In powder metallurgy, the process starts by mixing aluminum alloy powder with reinforcement particles in a uniform way. This mixture is then pressed under high pressure to form a compact shape and heated (sintered) so that the particles bond strongly as well as form a dense structure. This method enables precise control over reinforcement content, particle distribution, and microstructural refinement, allowing the development of composites with tailored strength, hardness, and wear resistance.

For applications where strength is needed in a specific direction and load-bearing capacity should be maximized, it is important to control the alignment and distribution of reinforcement particles during processing. By carefully selecting suitable reinforcements for the Al7068 matrix, engineers can produce components that are lightweight yet strong enough to withstand high mechanical stresses.

These specially designed Al7068 composites are highly useful for structural applications, where high strength, stiffness, and reliability are essential, while also offering the advantages of reduced weight and improved overall efficiency.

3.2 Fabrication and Characterization Techniques

i. **Al7068:** The process starts with the selection of Al7068 is one of the strongest and high strength aluminum alloys currently commercially available. It is widely use in the application of aerospace, defense, and high-performance structural applications. Al7068 have been selected due to its superior strength-to-weight ratio, excellent fatigue resistance, and improved resistance to stress corrosion compared to other conventional alloys such as Al7075. In the present research, Al7068 is used in the form of a solid billet, ensuring material homogeneity, dimensional stability, and consistent metallurgical characteristics for subsequent melting, processing, and composite fabrication.

ii. **Solid Billet:** The Al7068 alloy, provided by the supplier in the form of a solid billet, is initially cutting into smaller rectangular block to easy handling, melting, and uniform mixing of reinforcement. Precision methods such as CNC cutting, band sawing, or shearing are adopted for cutting operation.

The piecing of billets is essential for casting process because directly processing large pieces of Al7068 can lead to non-uniform thermal gradients during melting, which turns into micro segregation, porosity, or heterogeneous reinforcement distribution in composite fabrication. By reducing the billet size, the heat transfer efficiency and melt homogeneity are enhanced, thereby ensuring better wettability of ceramic reinforcements and also minimizing casting-related defects.

iii. **Casting:** For casting SiC and Si_3N_4 micro-particles are used together as reinforcements in the Al7068 matrix.

- **Micro SiC**: The fabricated castings were systematically categorized into five compositions with varying silicon carbide (SiC) reinforcement levels of 0%, 1.0%, 0.5%, 0.5 % and 1.0% by weight as shown in table 3.1. The addition of micro-scale SiC particulates have been deliberately chosen, its dimensional characteristics provide effective load transfer and enhanced interfacial bonding within the aluminum composite and improved mechanical performance, while simultaneously enabling microstructural refinement.
- **Micro Si_3N_4**: Analogous to the micro-SiC reinforced system, the present series of castings employed micro-scale silicon nitride (Si_3N_4) particulates as reinforcement, incorporated at varying volume fractions of 0%, 1.0%, 1.5%, 2.5%, and 2.0% as shown in table 3.1. The deliberate use of micro-sized Si_3N_4 particles will be intended

to enhance matrix-reinforcement interfacial bonding, promote microstructural homogeneity, and improve the resultant mechanical and tribological behavior of the composites.

Table 3.1. Sample wt 1600gm (Al7068)

Micro SiC	Micro Si_3N_4
0%	0%
1%	1%
0.5%	1.5%
0.5%	2.5%
1%	2%

iv. **Mechanical Testing**: After the casting process, the fabricated composites will be systematically characterized through a series of mechanical and microstructural testing. Mechanical performance will be assessed by using tensile, toughness, and hardness testing machine, while microstructural features will be examined by employing optical microscopy (OM) as well as by field emission scanning electron microscopy (FESEM) to elucidate particle distribution, interfacial integrity, and fracture morphology.

v. **Investigation Techniques:** Following the testing phase, four distinct analytical approaches will be employed, each providing complementary insights into the mechanical and microstructural performance of the composites.

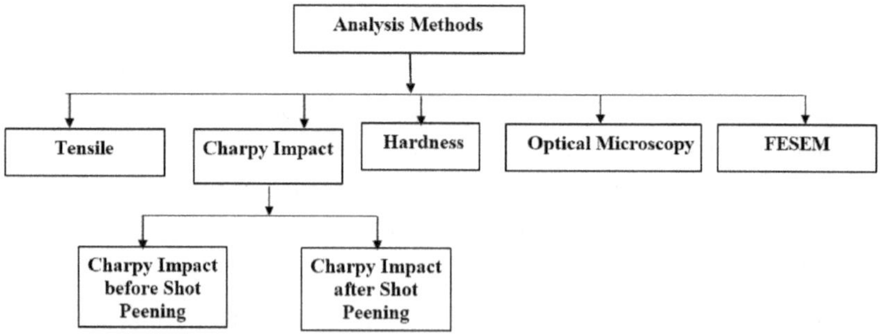

Figure 3.1 Analysis Approach

vi. **Tensile Testing**: Tensile testing is conducted to evaluate the ultimate tensile strength of a material for applications and to analyze its mechanical behavior when subjected to a tensile load and how much load it can withstand before failure. Tensile strength is defined as the maximum stress i.e. ultimate tensile strength of a material can endure under tensile load. So, it is a key parameter in material characterization and plays a crucial part in the design of structural and mechanical components.

In a typical tensile test, on UTM machine gradually increasing tensile load is applied on specimen until it breaks. The highest stress reached during this testing is gives the value of ultimate tensile strength, usually expressed in megapascals (MPa) or pounds per square inch (PSI). Apart from determining the maximum load-carrying capacity, tensile testing also provides important information about material behavior such as yield strength & elongation. Elongation of materials shows plastic deformation of aluiminum alloys before failure, i.e. it can stretch considerably while still carrying load. In contrast, brittle materials fail suddenly with very little plastic deformation before yielding.

The tensile strength mainly depends on many factors such as chemical composition, microstructural features, and treated temperature. Various post mechanical processes like alloying, T6 heat treatment process and work hardening also enhanced tensile properties, while polymers demonstrate strong temperature-dependent viscoelastic behavior.

Tensile testing helps the production and design engineers for optimize processes by predicting how materials will behave under mechanical stresses. In industries such as aerospace and automotive, tensile strength plays a key role in developing lightweight & strong components with improving safety, efficiency, and durability. Therefore, tensile strength is a critical mechanical parameter that ensures the reliability and suitability of materials for advanced engineering applications.

vii. **Rockwell Hardness Testing**: Rockwell hardness testing is performed to find the resistance to localized plastic deformation, such as indentation, scratching, or wear of the developed Al7068 composites to localized plastic deformation under standardized loading conditions. Compared to other techniques, the Rockwell method gives quick results, reliable assessment of bulk hardness, making it well suited for engineering-scale composite materials. In this hardness testing, a minor load is firstly applied by ball indenter on the specimen hold on anvil, followed by a major load using either a steel ball or a diamond indenter. The hardness value is determined directly from the depth of indentation or impression and expressed in the terms of Rockwell hardness number (HRB).

This method minimizes operator-related errors and ensures good repeatability, which is especially important when comparing composites containing different reinforcement proportions. To minimize microstructural non-uniformity variations, multiple hardness measurements is taken at different points on each specimen and averaged it all values. The recorded Rockwell hardness values is then correlated with tensile test results and microstructural observations, helping to clarify the role of reinforcement particles in enhancing the overall strengthening behavior of the Al7068 composites.

viii. **Charpy Impact Test**: Charpy impact testing is performed to evaluate the ability of Al7068-based composites to resist sudden impact loading under dynamic loading conditions. This test is particularly important for aerospace and automotive applications, where structural parts are sometimes exposed to sudden loads such as impacts and crash or collision situations. This result gives the clear information of the material's ability to withstand shock loading without catastrophic failure.

In this test, a notched specimen got fractured by swinging pendulum hammer, and the energy absorbed during fracture is measured in joules. By comparing the absorbed energy values for composites with different reinforcement compositions, the influence of micro and nanoparticles addition on ductility and brittleness properties was systematically evaluated.

Fracture behavior is further examined by using FESEM to get the clearer understanding of the failure mechanisms. The shot-peened samples showed more ductile characteristics, such as fine and deep dimples along with slower crack propagation. On the other hand, the untreated or as-cast specimens is mainly exhibited brittle or mixed-mode fracture features. These findings suggest that the combined effect of reinforcement and shot peening plays a significant role in improving the impact toughness of Al7068-based composites.

ix. **Optical Microscopy**: Researchers can better understand the microstructural behavior of the Al7068 alloy reinforced with micro or nano particles by using optical microscope testing. Its principle based on visible light and a system of lenses, including objective and eyepiece lenses, to magnify images features that are not visible by the naked eye, typically up to about 1000×. The resolving ability of the microscope depends on the wavelength of light as well as on numerical aperture of the objective lens, which together show the clarity of fine details.

Before testing, firstly samples are carefully prepared according to standard metallographic procedures including steps such as sectioning, grinding, polishing, and chemically etching. Proper surface preparation is required for optical microscopic to get true microstructural details such as grain boundaries, phase distribution, and reinforcement interfaces. It helps to corelate the relationship between manufacturing conditions such as casting parameters, reinforcement proportion, and post-treatment methods and the resulting mechanical properties such as hardness, tensile strength, and impact toughness.

x. **Field Emission Scanning Electron Microscopy (FESEM)**: Field Scanning Electron Microscopy (FESEM) is used to study casting composites reinforced with micro and nano particles in very detailed manner, especially its surface and internal microstructure. It is more advanced than optical microscopy because it uses a focused beam of electrons in place of light. This allows FESEM to produce highly magnified images (up to more than 10^5 times) with better depth and clarity. The images provide surface behavior, material composition, and crystal structure, making it easier to identify particle–matrix bonding and small defects.

In this study, secondary electron (SE) and backscattered electron (BSE) modes is used for observing reinforcement distribution, grain boundaries, and fracture surfaces as well as highlights each different feature.

FESEM shows very fine details of the material as well as how the reinforcement particles are spread throughout the aluminium alloy and bonding within the alloy, and whether there are small defects like pores, cracks, or particle clusters. The uniform distribution of reinforcement and strong interfacial bonding generally show the improvement in mechanical properties as well as clustering of particles or the presence of defects can reduce strength and durability.

3.3 Methodology

In the present study, Al7068 alloy have been selected as the base material due to its excellent strength-to-weight ratio and thermal resistance, while reinforcement such as micro particles is introduced to further enhance its properties.

The chemical composition of Al7068, as shown in Table 3.1, highlights the percentage of zinc, copper, magnesium, and trace elements in optimizing strength, toughness, and corrosion resistance, suitable for high-performance engineering applications such as aerospace, defense, and high-performance engineering applications.

This Al7068 aluminum alloy mainly consists of zinc (8.16 wt%), which is the major strengthening element in the 7000 series alloys, contributing significantly to precipitation hardening. Copper (1.82 wt%) and magnesium (2.38 wt%) further improve the aluminum alloy's strength through the formation of intermetallic phases, while chromium (0.006 wt%) refines grain structure and improves corrosion resistance. Trace elements such as titanium, zirconium, and beryllium act as grain refiners, improving structural uniformity. Minor percentage of elements such as iron, silicon, nickel, cobalt, vanadium, bismuth, lead, and tin are present in controlled amounts to balance machinability and mechanical performance.

With a purity of 99.99% and a melting point of 597°C, Al7068 showing superior thermal stability, making it highly suitable for aerospace, defense, and high-performance automotive applications where weight reduction is first priority without compromising strength. The synergy of these alloying elements provides Al7068 with its outstanding strength, durability, and wear resistance, distinguishing it from conventional aluminum alloys.

Table 3.2: Chemical composition of Al7068

Composites	Si	Fe	Cu	Mn	Mg	Cr	Zn	Ti	Pb	Sn	Ni	Co	V	Bi	Be	Zr
Wt%	0.070	0.200	1.820	0.011	2.380	0.006	8.160	0.036	0.040	0.030	0.006	0.020	0.010	0.030	0.0035	8.160
Purity	99.99%															
Melting Point	597°C															

The stir casting method is employed to fabricate Al7068-based hybrid metal matrix composites reinforced with silicon carbide (SiC) and silicon nitride (Si_3N_4).

The fixed amount of Al7068 alloy (1600 g) was taken as the matrix material for all samples, while varying proportions of SiC and Si_3N_4 reinforcements were introduced, as summarized in Table 3.2. The reinforcement content was adjusted in the range of 1–2.5 wt.% for Si_3N_4 and 0.5–1 wt.% for SiC.

Table 3.3: Reinforcement Proportions in Al7068 Alloy

Samples ID	Sample-Composition (wt% SiC + wt% Si_3N_4)	Al7068 (in grams)	SiC Weight (in grams)	Si_3N_4 Weight (in grams)
S0	Al7068/ (0.0% SiC+0.0%Si_3N_4)	1600	0	0
S1	Al7068/ (1% SiC+1%Si_3N_4)	1600	16	16
S2	Al7068/ (0.5% SiC+1.5%Si_3N_4)	1600	8	24
S3	Al7068/ (0.5% SiC+2.5%Si_3N_4)	1600	8	40
S4	Al7068/ (1% SiC+2%Si_3N_4)	1600	16	32

During processing, the Al7068 alloy is first heated above its melting point (~725°C). Pre-weighed quantities of SiC and Si_3N_4 is gradually added into the molten alloy as well as stirring is maintained at 1400 rpm for 15 minutes to ensure uniform particle distribution.

The molten composite is then poured into preheated molds and allowed to solidify under controlled conditions for nearly 10 hours.

i. **Phase-I, Material Procurement and Alloy Composition Analysis:** The alloy was procured from Nextgen Steel & Alloys, Mumbai, India, and its chemical composition was verified at Kanha Testing Labs Pvt. Ltd, Gurugram.

a

NEXTGEN STEEL & ALLOYS

TAX INVOICE

Description Of Goods: ALUMINIUM 7068

GRAND TOTAL: 33,630.00

GSTN: 27ATYPM3096M1ZP

Auth. Signatory

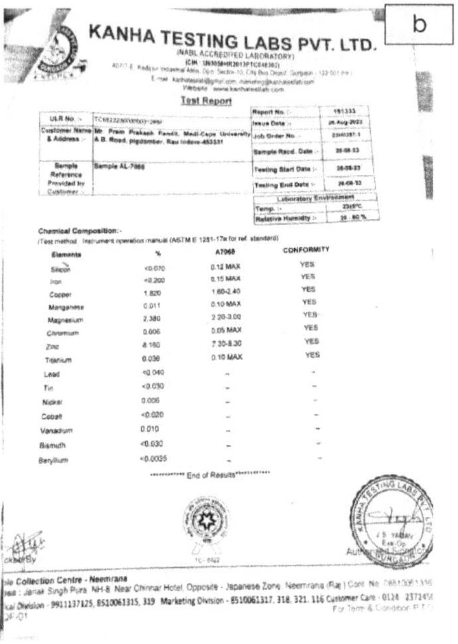

b

KANHA TESTING LABS PVT. LTD.

(NABL ACCREDITED LABORATORY)

Test Report

ULR No. :-		Report No. :-	151333
Customer Name & Address :-	Mr. Prem Prakash Pandit, Medi-Caps University A.B. Road, pigdamber, Rau Indore-453331	Issue Date :-	28-Aug-2023
		Job Order No. :-	
		Sample Recd. Date :-	25-08-23
Sample Reference Provided by Customer :-	Sample AL-7068	Testing Start Date :-	26-08-23
		Testing End Date :-	28-08-23

Laboratory Environment	
Temp. :-	23±5°C
Relative Humidity :-	30 - 60 %

Chemical Composition:-

(Test method : Instrument operation manual (ASTM E 1251-17a for ref. standard)

Elements	%	A7068	CONFORMITY
Silicon	<0.070	0.12 MAX	YES
Iron	<0.200	0.15 MAX	YES
Copper	1.820	1.60-2.40	YES
Manganese	0.011	0.10 MAX	YES
Magnesium	2.380	2.20-3.00	YES
Chromium	0.006	0.05 MAX	YES
Zinc	8.160	7.30-8.30	YES
Titanium	0.036	0.10 MAX	YES
Lead	<0.040	-	-
Tin	<0.030	-	-
Nickel	0.006	-	-
Cobalt	<0.020	-	-
Vanadium	0.010	-	-
Bismuth	<0.030	-	-
Beryllium	<0.0035	-	-

************ End of Results************

Authorised Signatory

Figure 3.2(a) Material Purchase Bill

Figure 3.2(b) Chemical Composition Test Report

Silicon carbide particulates generally ranging from 1-10 μm and a purity exceeding from 99 wt%. In contrast, Silicon nitride particulates are finer, typically 1-10 μm in size and purities between 97 and 99.8 wt% for composite applications as shown in Fig 3.3(b) and (c)

a

Figure 3.3(a) Al 7068

Figure 3.3(b) SiC (Size: 10 μm) Figure. 3.3(c) Si_3N_4 (Size: 10 μm)

ii. **Phase-II, Stir Casting:** Stir casting is one of the most commonly used liquid-state fabrication techniques for producing metal matrix composites (MMCs) because of its simplicity, cost-effectiveness, and suitability for large-scale production. In the present work, Al7068 alloy was selected as the matrix material and melted in a crucible at a temperature range of 700–800 °C, where it was held for about two hours to ensure complete melting, as illustrated in Fig. 3.4.

Once the alloy reaches a molten state, a pre-calculated composition of silicon carbide (SiC) and silicon nitride (Si_3N_4) reinforcement particles, as detailed in Table 3.2, were gradually mixed into the molten metal. To achieve uniform dispersion of the reinforcement particles and minimize agglomeration, mechanical stirring was carried out at a constant speed of 250 rpm for a duration of 15–20 minutes. Following thorough mixing, the composite melt was poured into a mold cavity and allowed to solidify, forming the final hybrid composite structure.

This method ensures significant improvements in the mechanical properties of the developed material, enhancing its tensile strength, wear resistance, and overall structural integrity for advanced engineering applications.

The processing parameters were selected based on literature and trial experiments. Proper melting temperature, stirring speed, stirring time, and reinforcement preheating were used to achieve uniform particle distribution, reduce porosity, and improve the mechanical properties of the developed composite.

The process can be delineated into the following sequential stages

a) **Melting the Matrix Material**:

- Batchwise balanced the weight of Al7068 and the composition of the particles SiC & Si_3N_4 has been prepared. The process begins with the melting of the base matrix alloy, typically aluminum, magnesium, or copper alloys, in a controlled furnace environment.
- The Al7068 alloy has been placed in a crucible and heated in an electric furnace for approximately two hours at a temperature range of 700–800°C, surpassing its melting point to achieve a fully molten state. The melt temperature has been carefully regulated to achieve complete liquefaction while minimizing oxidation and other undesirable chemical reactions through the use of protective atmospheres or fluxes.
- Flux degassing agents (hexachloroethane) as shown in fig is mixed to the molten metal to eliminate impurities and improve the casting quality.

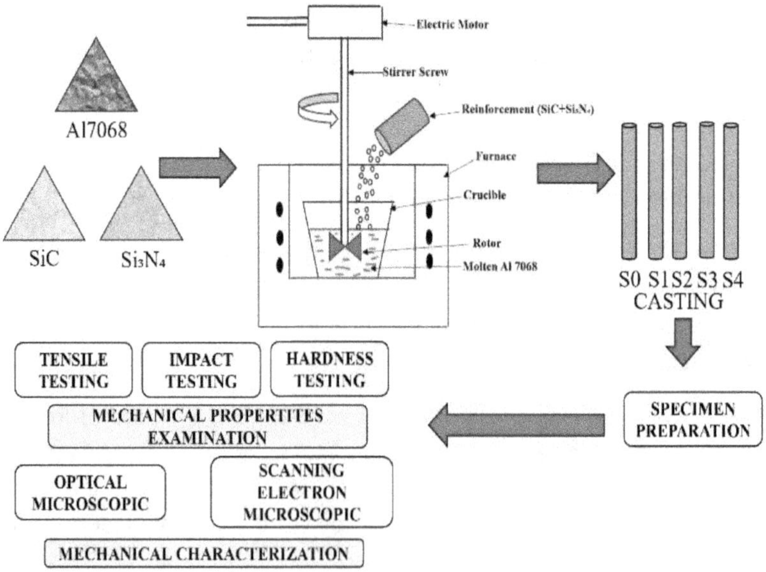

Figure. 3.4 Flowchart of the research to be carried out

b) **Adding Reinforcements**:

- Pre-measured amounts of SiC and Si_3N_4 microparticles (refer Table 3.3) were gradually introduced into the molten metal.

c) **Stirring Process**:

- The mixture was stirred at 250 rpm for 15–20 minutes to ensure a uniform dispersion of reinforcements throughout the matrix.

d) **Casting**:

- Once a uniform dispersion of reinforcement was achieved, the composite melt was transferred into preheated molds of the desired geometry.
- The pouring process were performed under controlled conditions to minimize turbulence and suppress gas entrapment, thereby reducing porosity in the solidified product.

e) **Solidification and Cooling**:

- The molten alloy was maintained in a stationary state and subsequently poured into a preheated metallic mold to obtain the desired castings.

f) **Post-Processing**:

- The solidified MMCs may undergo further post-processing treatments such as machining, heat treatment, and surface modification.
- These operations are employed to refine the microstructure, enhance mechanical performance, and achieve the dimensional and surface finish requirements essential for aerospace, automotive, and structural applications.

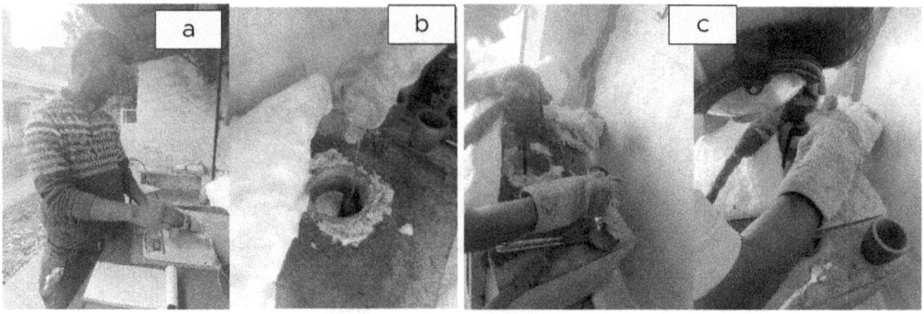

Figure 3.5 (a) Weighing of Al7068 (b) Removal of Slag (c) Stiring & Mixing (Micro Particles) of Molten metal

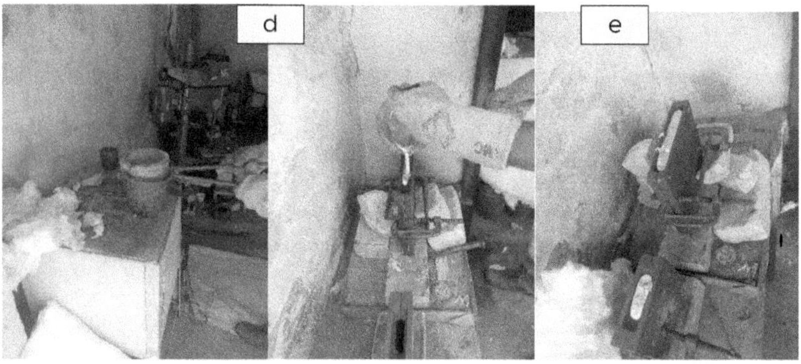

Figure 3.5 (d) Crucible kept out from furnace (e) Pouring into Pattern

Figure 3.5 (f) Desired casting

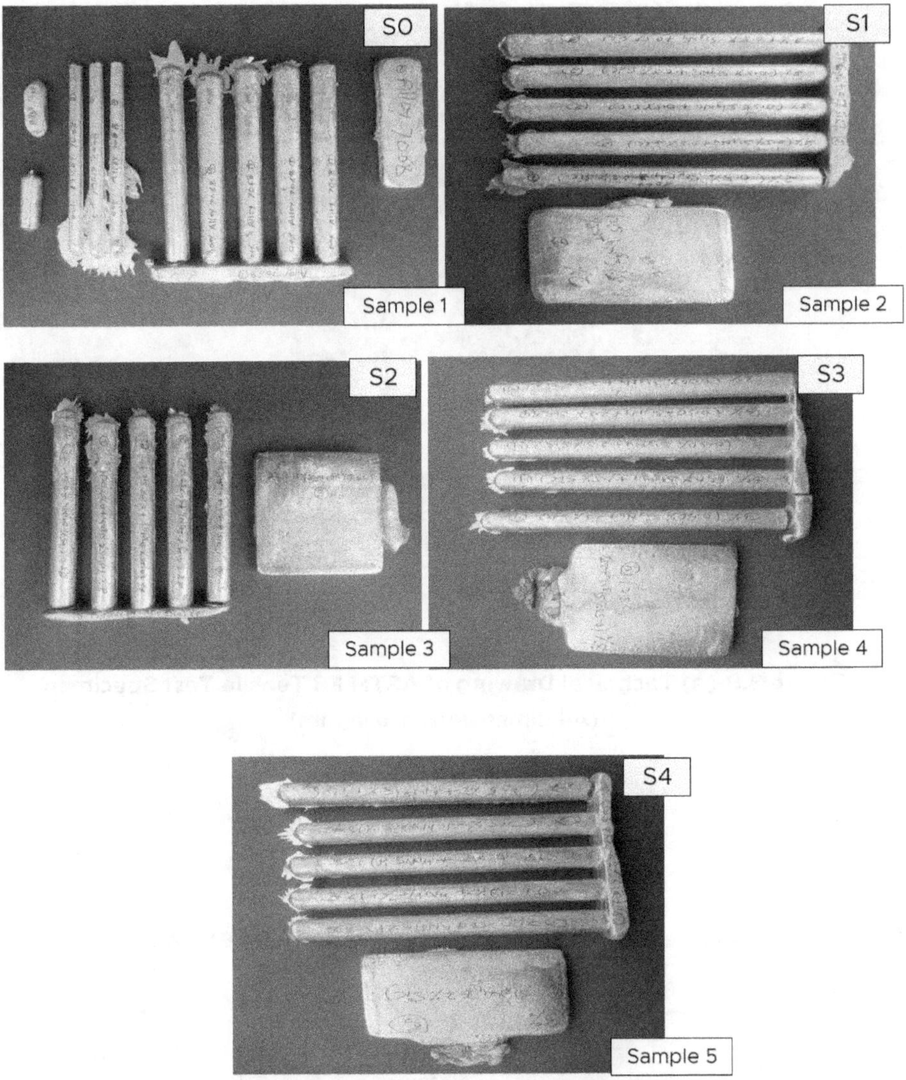

Figure 3.6 Final Casting Al7068 with different compositions

3.4 Preparation of Standard (ASTM) Specimen

Figure 3.7 Preparation of Tensile specimen

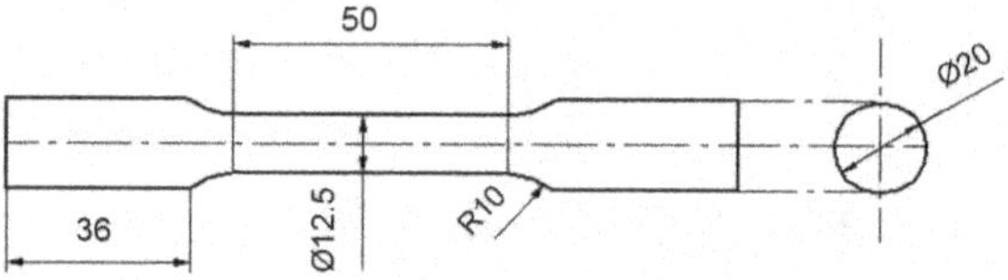

Figure 3.8 (a) Technical Drawing of ASTM E8 Tensile Test Specimen (All dimensions are in mm)

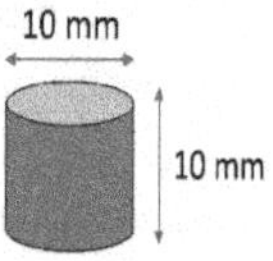

Figure 3.8 (b) Technical Drawing of ASTM E18 Pin

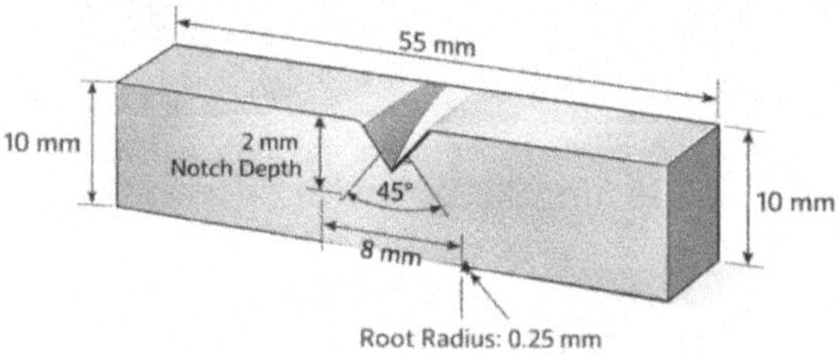

Figure 3.8 (c) Technical Drawing of ASTM E23 Charpy test specimens

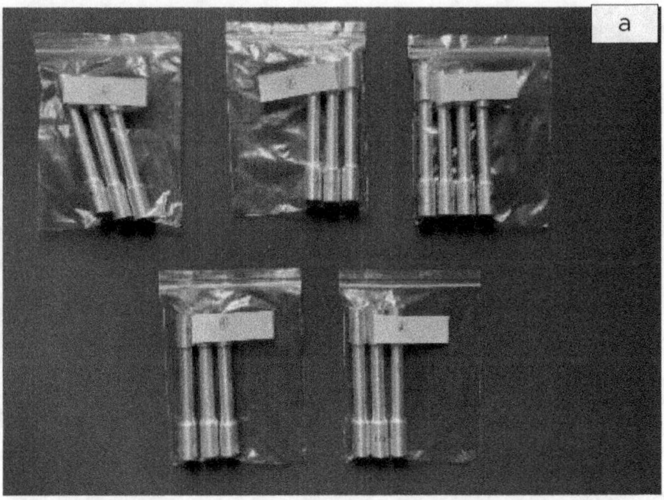

Figure 3.9(a) Tensile Test Specimen

Figure 3.9(b) Impact Test Specimen

C

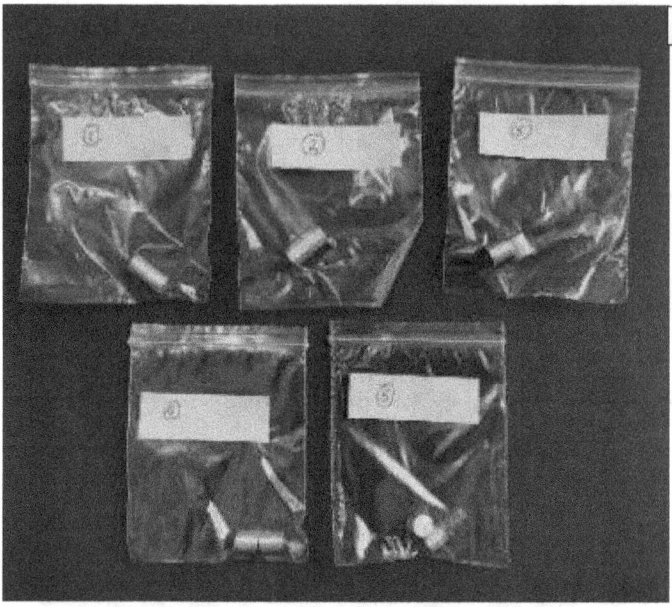

Figure 3.9(c) Rockwell Hardness Test Specimen

3.5 Proposed Approach for Alloy Synthesis

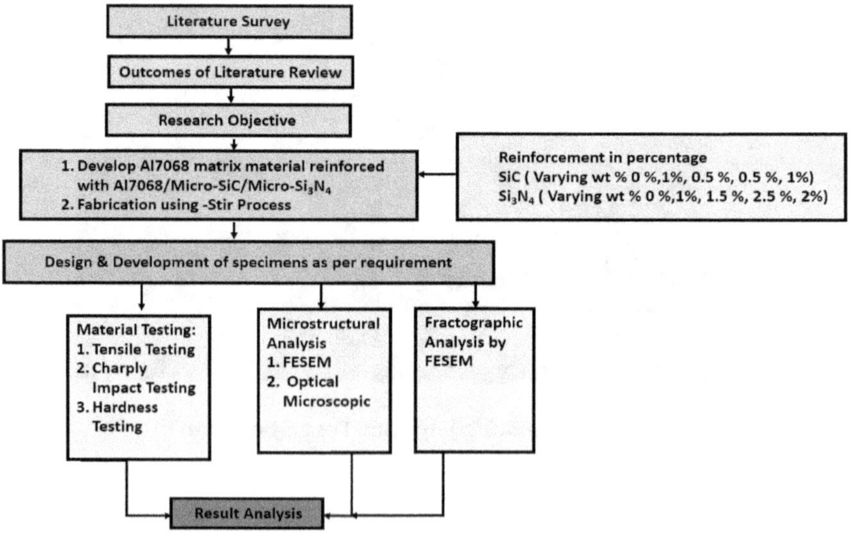

Figure 3.10 Proposed alloy preparation process with micro SiC & Si_3N_4

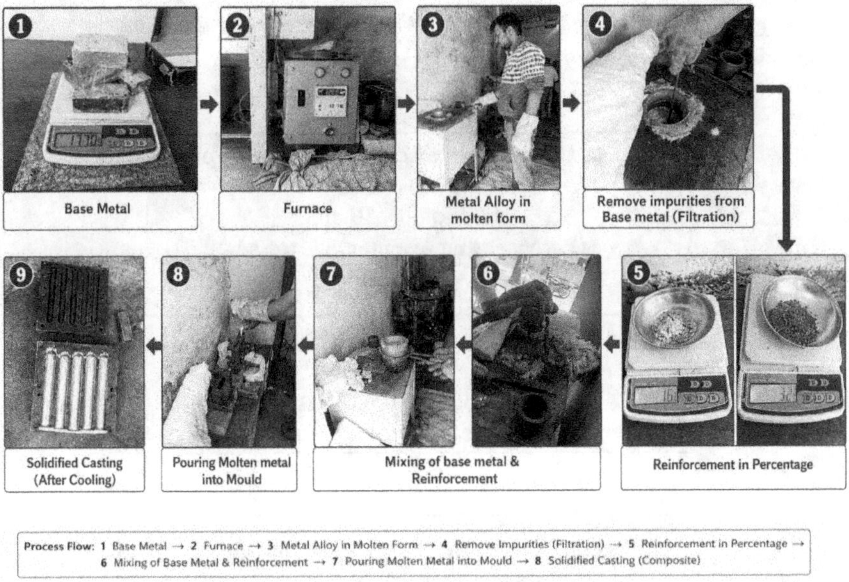

Figure 3.11 Alloy preparation process with micro SiC & Si_3N_4

3.6 T6 Heat Treatment

The T6 heat treatment process comprising of solution heat treatment at 500–540°C followed by water quenching and artificial ageing at 120–130°C for 6 hours is the most effective heat treatment for Al 7068 hybrid composites.

This heat treatment process improves tensile strength and hardness due to the formation of fine strengthening precipitates and improved matrix-reinforcement bonding.

Figure 3.12 T6 Heat Treatment Process

Chapter - 4
Results and Discussion

Chapter 4
Results and Discussion

This chapter presents a detailed analysis and discussion of the experimental results obtained during these research investigations, in conformity with the stated objectives of the thesis. For greater clarity and systematic understanding, the findings have been organized into well-defined sections, each addressing specific aspects of the study and its corresponding interpretations.

a. **Mechanical Properties of Al-SiC/Si_3N_4 Alloys:** This section examines the mechanical properties of Al-SiC/Si_3N_4 alloys, including mechanical parameters like tensile strength, hardness, impact strength are systematically evaluated before shot peening & after shot peening in order to quantify the effect of surface treatment on energy absorption and fracture resistance, all data evaluated at room temperature.

b. **Characterization of Al- SiC/Si_3N_4 Alloys by optical scanning:** This section covers the microstructural evaluation of Al-SiC/Si_3N_4 composites, containing different compositions of micro silicon carbide & silicon nitrite is carried out using optical microscopy and FESEM at different magnifications.

The results presentation and discussion are systematically explained for each material composition and under varying resolution conditions.

4.1 Investigation of Mechanical Properties

To evaluate the mechanical properties of Al7068/SiC/Si_3N_4 composites such as-

- Tensile Testing including Ultimate Tensile Strength, Yield Strength and Elongation.
- Impact Strength before and after Shot Peening
- Rockwell Hardness Testing

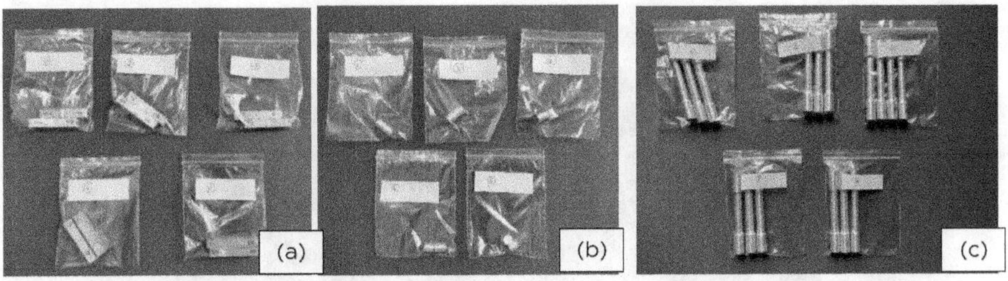

Figure 4.1 (a) Charpy Test Specimen (b)Rockwell Hardness Test Specimen (c) Tensile Test Specimen

4.1.1 Investigation of Tensile Test

Tensile testing was carried out using a Universal Testing Machine (Type of Machine: Hydraulic, Max Capacity: 100 kN-600 kN) and conducted at room temperature at CSIR, Durgapur.

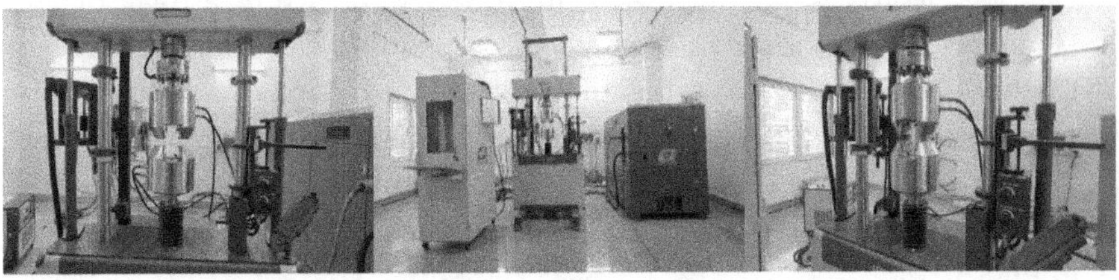

Figure 4.2 Universal Testing Machine

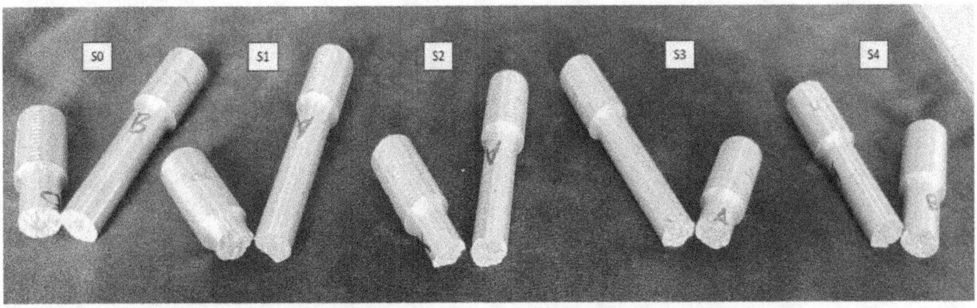

Figure 4.3 (a) Tensile Test Specimen After Test

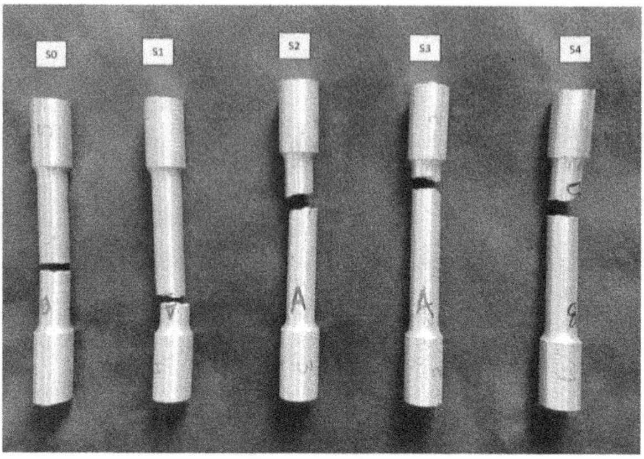

Figure 4.3 (b) Tensile Test Specimen After Test

Figure 4.4 (a) The chart displays the Ultimate Tensile Strength (UTS) of samples S0 to S4. UTS increases from 635 MPa (S0) to a peak of 752 MPa in S4. The highest improvement was observed in S4 (18.43%) relative to the baseline S0. indicates that it has the optimum reinforcement composition and/or processing condition. while S0 exhibited the weakest mechanical performance. The variations indicate differences in material properties, potentially due to composition, processing, or heat treatment effects.

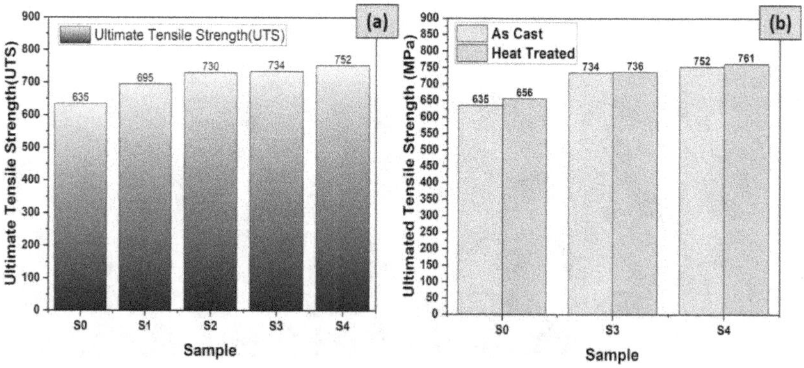

Figure 4.4 (a) Ultimate Tensile Strength (b) Ultimate Tensile Strength: As-Cast vs. Heat-Treated Condition

Figure 4.4 (b) compares the ultimate tensile strength (UTS) of samples S0, S3, and S4 in both "As Cast" and "Heat Treated" conditions. Heat treatment enhanced UTS in all cases: S0 increases from 635 MPa to 656 MPa, S3 from 734 MPa to 736 MPa, and S4 from 752 MPa to 761 MPa. This demonstrates that heat treatment improved mechanical strength by refining microstructure and reducing internal stresses, making materials more suitable for demanding structural applications.

Figure 4.4 (c) shows the Yield Strength (YST) of samples S0 to S4. YST increases from 310 MPa in S0 to a maximum of 374 MPa in S4. This trend indicates that sample S4 has the highest resistance to plastic deformation, making it the strongest among all. The highest improvement is observed in S4 (20.65%) as compared to S3 (17.74%) as shown in fig 4.4 (g) as well as S4 demonstrated optimal mechanical properties for structural applications.

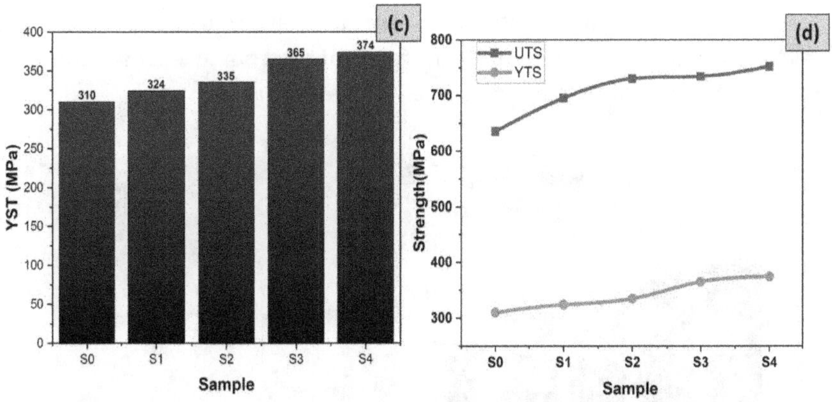

Figure 4.4 (c) Yield Tensile Strength (d) Compare UTS and YTS

Figure 4.4 (d) shows a comparison of Ultimate Tensile Strength (UTS) and Yield Strength (YTS) across all samples from S0 to S4. Both strengths reached peak at sample S4, where the UTS reached approximately 752 MPa and the YTS around 374 MPa, indicating the highest mechanical performance. UTS consistently trends higher than YTS, as expected. Among all compositions S4 has the highest total reinforcement (3%) with balanced material composition and demonstrates the highest UTS and YTS, signifying maximum strengthening effect.

Figure 4.4 (e) shows the percentage elongation of five samples (S0–S4), which indicate the ductility of the material. The elongation decreases from 15.2% for sample S0 to 9.35% for sample S4. Sample S4 exhibits the lowest elongation among all samples. This decreasing trend suggests that variations in composition or processing have enhanced the mechanical performance of the material. The reduction in elongation indicates an increase in strength, highlighting an inverse relationship between ductility and strength, which is beneficial for material optimization.

Figure 4.4 (f) compares the % elongation of "As Cast" and "Heat Treated" samples (S0, S3, S4). compares the percentage elongation of "as-cast" and "heat-treated" samples (S0, S3, and S4). Heat treatment consistently reduces elongation, with S0 decreasing from 15.2% to 13.8%, S3 from 11.6% to 10.2%, and S4 from 9.35% to 9.1%. This reduction indicates an improvement in ultimate tensile strength and overall mechanical performance after heat treatment. The slight decrease in elongation following T6 heat treatment can be attributed to improved microstructural uniformity and the relief of internal stresses during the solution treatment process, which enhances strength.

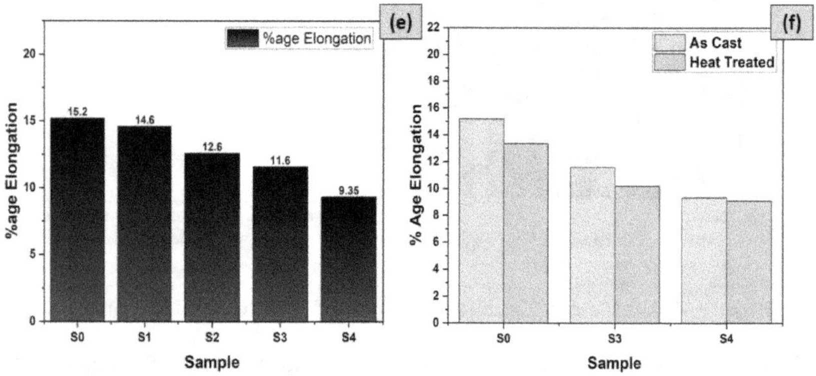

Figure 4.4 (e) Percentage Elongation (f) Percentage Elongation: As-Cast vs Heat Treated Condition

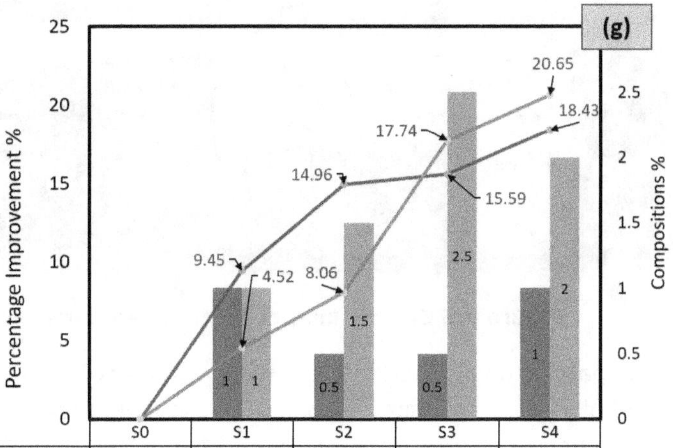

	S0	S1	S2	S3	S4
Silicon Carbide	0	1	0.5	0.5	1
Silicon Nitride	0	1	1.5	2.5	2
UTS (%)	0	9.45	14.96	15.59	18.43
YTS (%)	0	4.52	8.06	17.74	20.65

Figure 4.4 (g) Percentage Improvement of UTS and YTS Reference to Base Alloy Al7068

4.1.2 Charpy Impact Testing before shot Peening

The Charpy testing was done by means of a Digital Impact Testing Machine (Machine Model: SE-IT 300, SR NO: 503/17-18), Max Capacity: - 300 J) in room temperature at Material Testing Laboratory, MANIT, Bhopal.

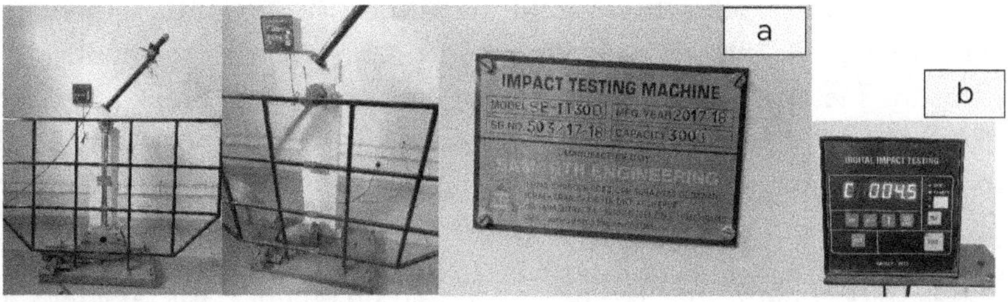

Figure.4.5 (a) Digital Impact Testing Machine (b) Air Resistance

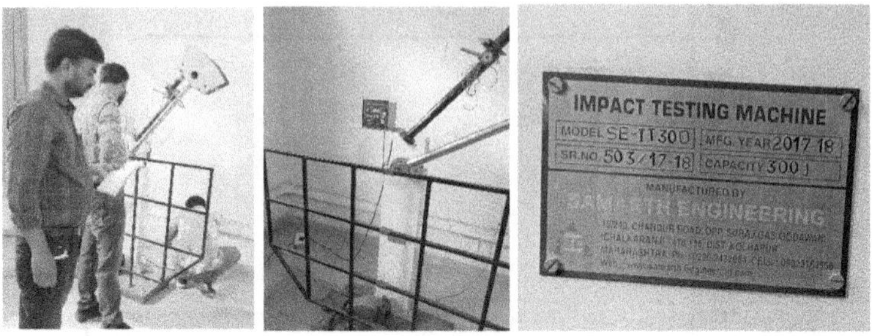

Figure 4.6 Charpy Imapct Testing Readings

Charpy impact testing is used to study the toughness of the Al7068 alloy and its reinforced composite samples (S0–S4). The amount of energy absorbed during fracture, expressed as net heat absorbed in Joules indicates how well the material can resist sudden impact or shock loading. The base Al7068 alloy (S0) exhibited the lowest impact energy of 1.0 J and therefore was taken as the reference material.

The base Al7068 alloy (S0) had the lowest impact energy of 1.0 J and was used as the reference for comparison. When reinforcement was added, the toughness improved in all the composite samples from S0 to S4.

Sample S1 absorbed 1.6 J of energy, showing a 60% increase over the base alloy. Sample S2 reached 1.4 J, which is a 40% improvement, while sample S3 showed a smaller rise to 1.2 J, indicating a 20% increase in toughness.

Among all the samples, S4 performed the best with an impact energy of 2.0 J, which is double that of the base alloy (100% improvement) as shown in fig.4.8. This shows that the combination of reinforcements in S4 helps the material absorb more energy and slows down crack formation during impact. These significant increase in impact energy suggests that the reinforcement combination in S4 effectively enhances energy absorption and delays crack propagation during impact loading.

The results confirm that the mixing of reinforcement particles plays an important role in improving the impact resistance of the Al7068 matrix. In particular, sample S4 exhibited the most favorable toughness characteristics, making it a promising material for applications where resistance to sudden impact and dynamic loads is required.

The improvement trend supports the hypothesis that suitable reinforcement selection and distribution within the aluminum matrix play a crucial role in resisting crack propagation and increasing energy absorption capacity.

Table 4.1. Charpy Impact Testing Reading before Shot Peening

Sample	Energy Absorbed (J)	Air Resistance & Other losses (J)	Net Energy Absorbed (J)
S0	5.5	4.5	1
S1	6.1	4.5	1.6
S2	5.9	4.5	1.4
S3	5.7	4.5	1.2
S4	6.5	4.5	2

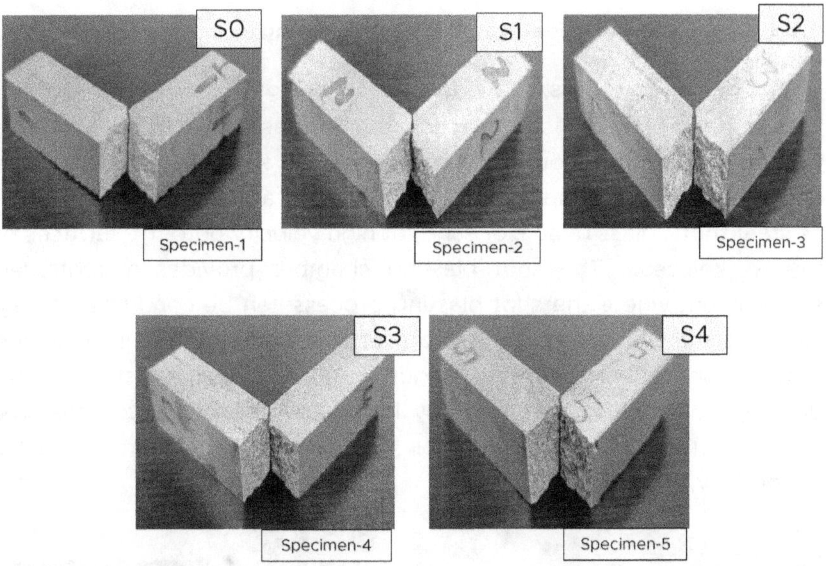

Figure 4.7. Specimen After Charpy Test Before Shot Peening

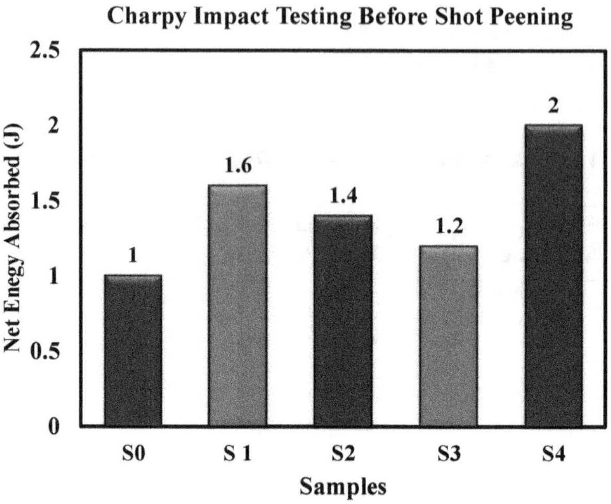

Figure 4.8 Variation in Impact Value for Al7068 and its Composites

4.1.3 Charpy Impact Testing after shot Peening

In shot peening process, all the five specimens of Al7068/SiC/Si_3N_4 kept together and placed in shot blasting chamber as shown in fig. 4.10. This treatment was carried out to clean, strengthen, and improve the surface characteristics of the specimens by propelling abrasive materials (such as steel shots, glass beads, or sand) at high velocity onto the surfaces of the workpieces. The shot blasting chamber provides a controlled environment where the shot blasting process can be conducted safely and efficiently. Shot peening is a powerful surface treatment technique that is use to enhance the fatigue strength, wear resistance, and durability of metal components by introducing beneficial compressive residual stresses. This operation performed at Material Testing Laboratory, MANIT, Bhopal.

Figure 4.9 Shot Blasting Chamber

Figure 4.10 Specimen's placed in the Shot Blasting Chamber

After shot blasting process, one by one Charpy impact testing performed on the specimen to found the impact value as shown in Table 3.

Table 4.2 Charpy Impact Testing Reading after shot peening

Sample	Energy Absorbed (J)	Air Resistance (J) & Other losses (J)	Net Energy Absorbed (J)
S0	5.7	4.5	1.2
S1	6.5	4.5	2
S2	6.0	4.5	1.5
S3	5.9	4.5	1.4
S4	7.1	4.5	2.6

Figure 4.11 Specimen after Charpy Test after shot peening

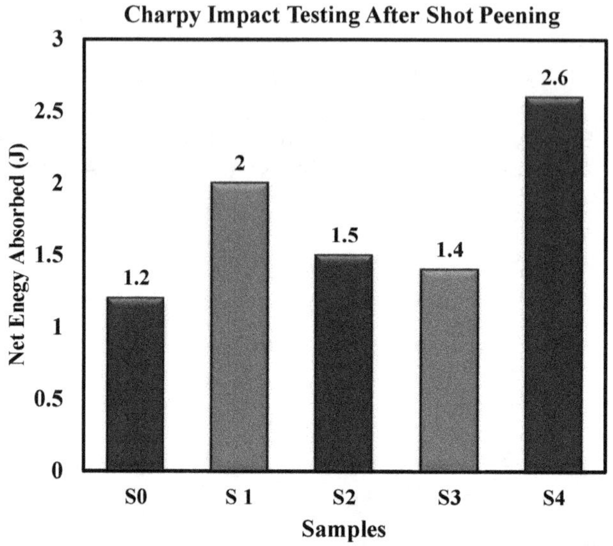

Figure 4.12 Variation in Impact Value for Al7068 and its Composites After Shot Peening

The Charpy impact testing have been performed on five samples (S0 to S4) to evaluate the improvement in impact resistance following shot peening treatment. After shot peening operation, Charpy impact testing was conducted on all five samples to assess the improvement in impact resistance. Sample S0, the unreinforced base alloy, absorbed 1.2 J of impact energy. After shot peening, all the composite samples showed better toughness. Sample S1 absorbed 2.0 J, which is about a 66.67% increase compared to S0. Sample S2 showed a moderate improvement with 1.5 J (about 25% higher than the base alloy) as well as sample S3 absorbed 1.4 J, giving a 16.67% improvement.

Among all the samples, S4 gave the best result by absorbing 2.6 J of energy as shown in fig. 4.12. This is a significant increase of about 116.67% compared to the base alloy, showing that it has the highest ability to withstand impact after treatment. The shot peening process clearly improves the impact resistance of the material with S4 showing the highest improvement. Comparison of Charpy impact test before and after shot peening as shown in fig. 4.13.

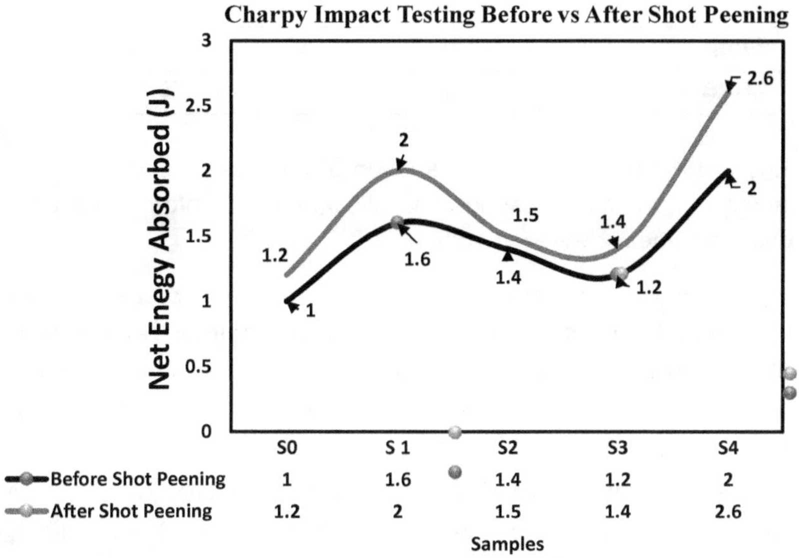

Figure 4.13 Comparison Graph of Charpy Impact Strength before and After Shot Peening

4.1.4 Rockwell Hardness Testing

The Rockwell Hardness test was done by means of a Hardness Tester (Machine Model: SSS- RASN (E)(B), SR NO: 2023-24//208) in room temperature at Material Testing Laboratory, MANIT, Bhopal.

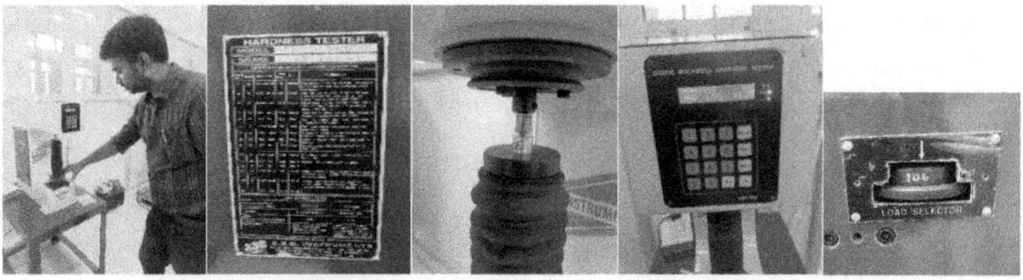

Figure 4.14 Rockwell Hardness Testing Machine

From the graph it is observed that the hardness increases with increase in weight percentage of reinforcement, this indicates good bonding between base alloy and reinforcement; shows the hard particles are successfully dispersed in the matrix

Table 4.3 Rockwell Hardness Readings of Al7068 and its composites

Sample	S0	S1	S2	S3	S4
HRB	42.4	57.5	56	59.7	65.57

Figure 4.15 below shows the variation of the hardness with percentage increase in the silicon carbide and silicon nitride reinforcement particles under non-heat-treated condition.

Figure 4.15 illustrates the hardness of the sample as cast and after heat treatment. The base Al7068 alloy (S0) had a hardness of 42.4 HRB. When silicon nitride (Si_3N_4) and silicon carbide (SiC) were added, the hardness increased in all the samples. Sample S1 (1% SiC + 1% Si_3N_4) showed a hardness of 57.5 HRB, which is about a 35.61% increase compared to the base alloy. Sample S2 (0.5% SiC + 1.5% Si_3N_4) reached 56.0 HRB, giving a 32.08% improvement. Sample S3 (0.5% SiC + 2.5% Si_3N_4) showed a higher hardness of 59.7 HRB, which is about a 40.80% increase. The highest hardness was seen in Sample S4 (1.0% SiC + 2.0% Si_3N_4), with a value of 65.57 HRB, showing a strong improvement of 54.61% over the base alloy.

Among all the samples, the heat-treated S4 showed the highest hardness value of 104.6 HRB. This clearly shows that combining different reinforcements along with heat treatment gives much better mechanical properties.

This progressive increase in hardness indicates the effective role of ceramic reinforcements in enhancing the mechanical strength of Al7068. These particles make the internal structure finer and act as barriers to dislocation movement, which helps the material carry more load and improve load-bearing capacity.

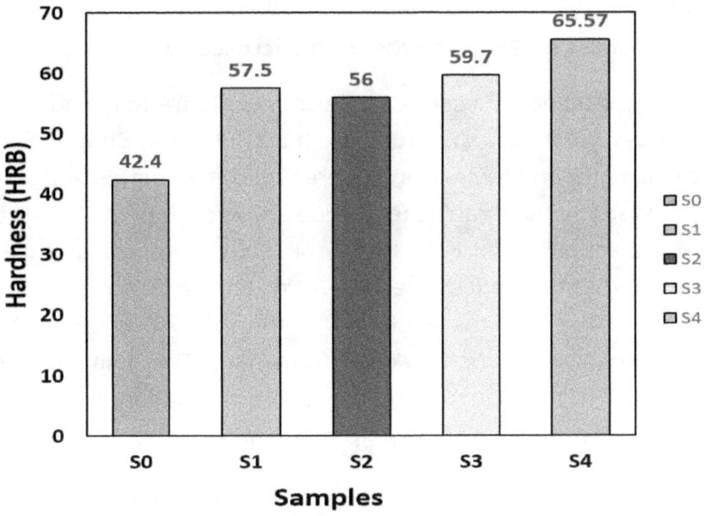

Figure 4.15 Variation in HRB for Al7068 and its Composites

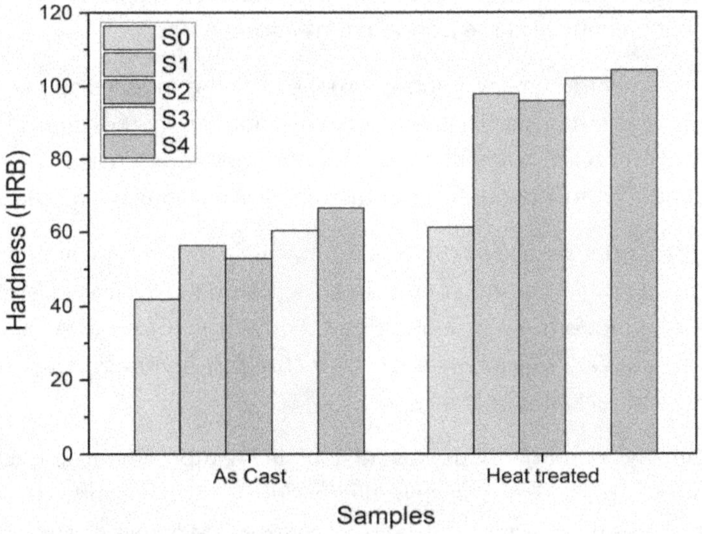

Figure 4.16 Comparison of Hardness (HRB) for Al7068 (As Cast vs Heat treated)

4.2 Material Characterization

4.2.1 Field Emission Scanning Electron Microscopy

FESEM were studied by using Carl Zeiss make ultra Plus coupled with energy dispersive X-ray. spectroscopy (EDX) at IISER, Bhopal. Samples of 10mm diameter and 10mm length were cut, cold mounted and polished as per standard metallographic technique and etched by Keller's reagent (3 ml HCl, 2 ml HF, 5 ml HNO3, 190 ml H2O). The scanning electron microscopy (SEM) images presented in Fig. 4.17 (a–e) (at 300× magnification) illustrate the evolution of the microstructure in Al7068 alloy with varying levels of hybrid reinforcement. These images provide clear insights into the matrix morphology, grain boundary networks, and dispersion of reinforcements such as Si_3N_4 and SiC.

The unreinforced Al7068 alloy (Figure 17a) displays a typical cast dendritic microstructure with prominent, interconnected aluminum dendrites. Minimal refinement is observed, and the microstructure shows solute segregation along the grain boundaries, indicative of limited solidification control in the absence of reinforcement.

With initial reinforcement, (Figure17 b) the grain boundaries appear more refined, and small particulates begin to appear in the interdendritic regions. The reinforcement initiates heterogeneous nucleation, which disrupts the coarse dendritic structure and encourages grain refinement.

The microstructure shows (Figure 17c) further refinement with a notable reduction in dendrite arm coarseness. Reinforcement particles are more visible and moderately well-dispersed. Early signs of particle clustering begin to appear, which, while locally increasing hardness, may lead to microstructural inhomogeneity.

A relatively homogeneous dispersion of reinforcement particles is observed (Figure 17d). The dendritic network is significantly disrupted, and the grain structure appears equiaxed. Minimal clustering and enhanced matrix-reinforcement bonding suggest improved load transfer and resistance to plastic deformation, correlating well with the increased hardness observed experimentally.

At the highest reinforcement content, (Figure 17e) considerable clustering of particles is evident. Although the grain boundaries are still refined, the non-uniform distribution of reinforcements can act as stress concentrators. This condition may enhance hardness but could negatively impact ductility and fracture toughness due to localized embrittlement.

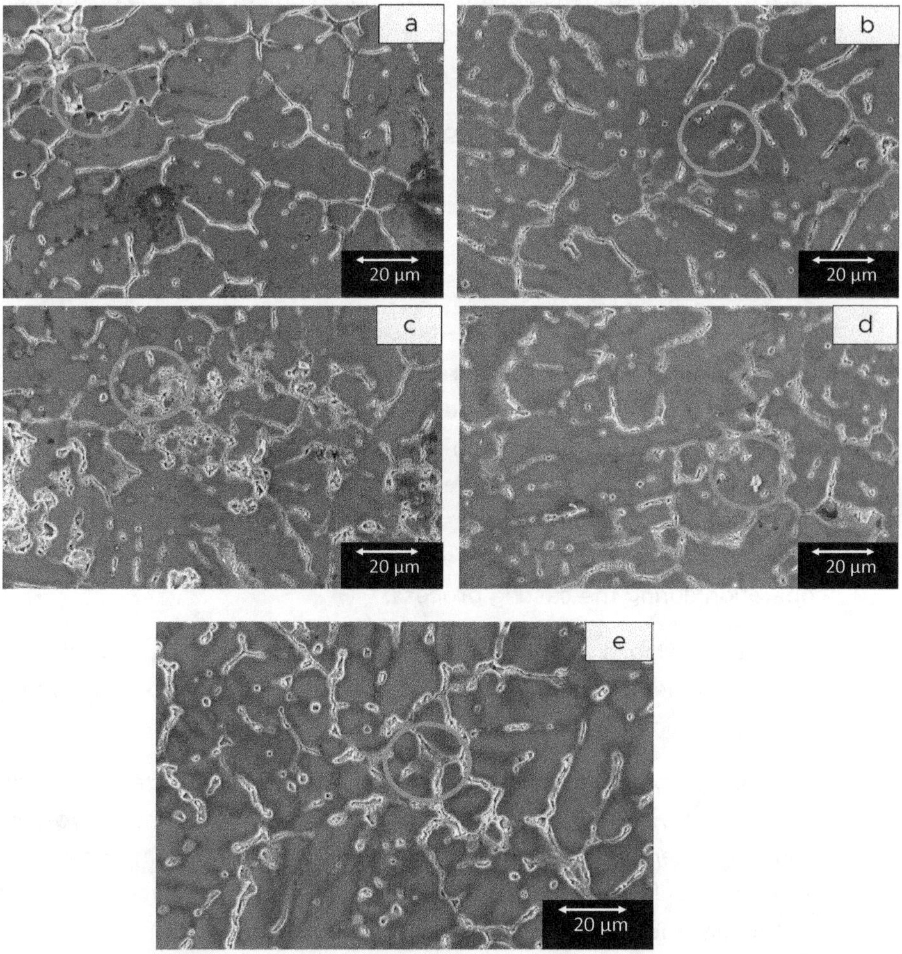

Figure 4.17 FESEM Images for Al 7068, 20 µm, Mag: 300X (a) Al7068 alloy [S0] (b) Al7068/(1% SiC +1 % Si_3N_4) [S1], (c) Al7068/(0.5% SiC +1.5% Si_3N_4) [S2], (d) Al7068/(0.5% SiC +2.5% Si_3N_4) [S3] & (e) Al7068/(1.0% SiC +2.0% Si_3N_4) [S4]

Figure 4.18 (Sample 1–Sample 5) presents scanning electron micrographs of the Al7068 alloy and its metal matrix composites at 500× magnification, the effect of reinforcement addition on microstructural refinement, micro particle silicon carbide & silicon nitride dispersion, and matrix continuity are highlighting. At this 500× magnification, finer details such as dendrite arm fragmentation, reinforcement particles distribution, and micro-defects are become clearly visible.

The micrograph of Sample 1 showing a coarse dendritic α-Al structure with thick and interconnected grain boundaries. The dendrite arms are irregularly arranged and widely spaced, indicating the unrestricted grain growth of casting during solidification of alloy. Without addition of reinforcement particles results in limited grain refinement, which is typically associated with lower strength and hardness.

With the mixing of less proportions of silicon carbide & silicon nitride micro particles, a noticeable refinement in the dendritic structure is observed in Sample 2. The dendrite arms become appeared thinner and less continuous compared with Sample 1. Small silicon carbide & silicon nitride micro particles are visible within the interdendritic regions, acting as heterogeneous nucleation sites and restricting grain growth. The distribution of particles is relatively uniform, indicating effective stirring operation during the casting process.

Sample 3 exhibits further refinement by increase the proportions of silicon carbide & silicon nitride micro particles of the matrix with a higher density of reinforcement particles.

The dendritic arms are significantly disrupted, and particle-rich regions are evident. However, localized clustering of particles and micro-voids can be observed. While such clustering can enhance hardness and strength locally, it may also introduce stress concentration sites that negatively affect ductility and fracture resistance.

This sample 4 shows one of the most uniform and refined microstructures compared to all compositions.

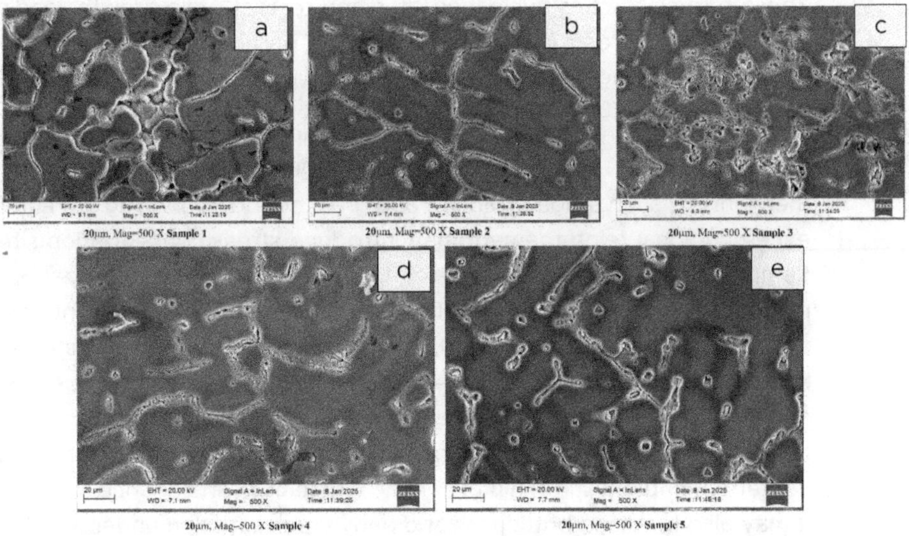

Figure 4.18 FESEM Images for Al 7068, 20 μm, Mag: 500X (a) Al7068 alloy [S0] (b) Al7068/(1% SiC +1 % Si_3N_4) [S1], (c) Al7068/(0.5% SiC +1.5% Si_3N_4) [S2], (d) Al7068/(0.5% SiC +2.5% Si_3N_4) [S3] & (e) Al7068/(1.0% SiC +2.0% Si_3N_4) [S4]

Figure 4.19 presents scanning electron micrographs of the Al7068 alloy and its hybrid composites observed at 700× magnification. This higher magnification image of SEM provides a clearer view of fine microstructural structures of Al7068 such as dendrite fragmentation, reinforcement–matrix interfaces, micro-voids, and particle clustering, providing detailed information about the effect of silicon carbide & silicon nitride micro particles addition on microstructural evolution.

The SEM image of Sample 1 shows a coarse and continuous dendritic α-Al matrix with thick grain boundaries. The dendrite arms are irregular and interconnected, indicating uncontrolled grain growth during solidification. The absence of reinforcement particles results in a relatively smooth matrix with limited barriers to dislocation motion, which indicated the comparatively lower strength and hardness of the base alloy. With the addition of silicon carbide & silicon nitride micro part reinforcement, Sample 2 exhibits noticeable refinement of the dendritic structure. The dendrite arms appear thinner and more segmented, and fine particles are observed along the grain boundaries and within interdendritic regions.

These particles act as nucleation sites and restrict grain growth, leading to improved microstructural uniformity. The particle dispersion is largely uniform, indicating effective stirring during fabrication.

The microstructure of Sample 3 reveals further disruption of the dendritic network. These micro structure of Al7068 showing strengthening mechanisms such as load transfer and grain boundary pinning are active, excessive local reinforcement may introduce stress concentration sites. Sample 4 demonstrates the most refined and homogeneous microstructure among all samples. The matrix–reinforcement interface appears clean and well bonded, indicating efficient stress transfer and improved resistance to plastic deformation. Minimal porosity is observed, suggesting an optimal balance between reinforcement content and processing conditions. At the maximum proportions of reinforcement content, Sample 5 still exhibits particle clustering can enhance hardness, it may also localized brittleness and early crack initiation under tensile or impact loading. Several micro particle-rich areas and isolated voids are visible.

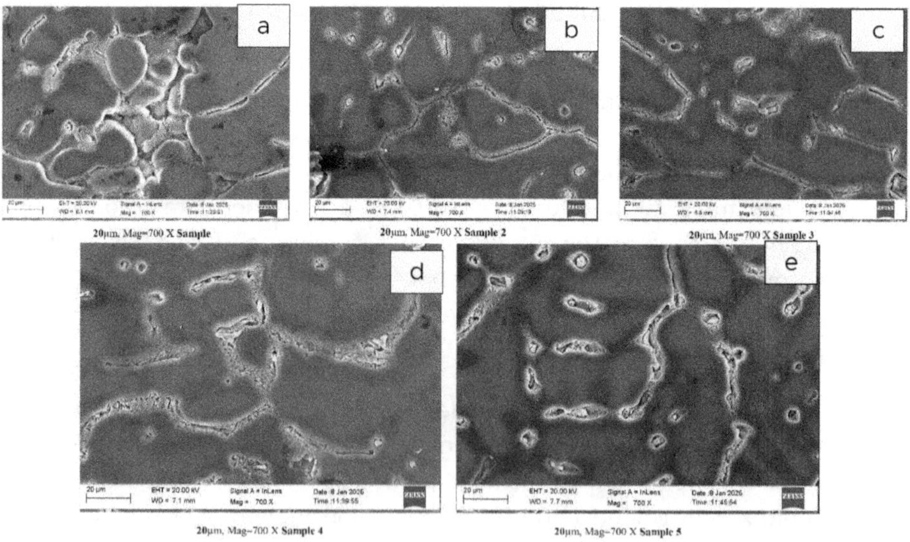

Figure 4.19 FESEM Images for Al 7068, 20 μm, Mag: 700X (a) Al7068 alloy [S0] (b) Al7068/(1% SiC +1 % Si_3N_4) [S1], (c) Al7068/(0.5% SiC +1.5% Si_3N_4) [S2], (d) Al7068/(0.5% SiC +2.5% Si_3N_4) [S3] & (e) Al7068/(1.0% SiC +2.0% Si_3N_4) [S4]

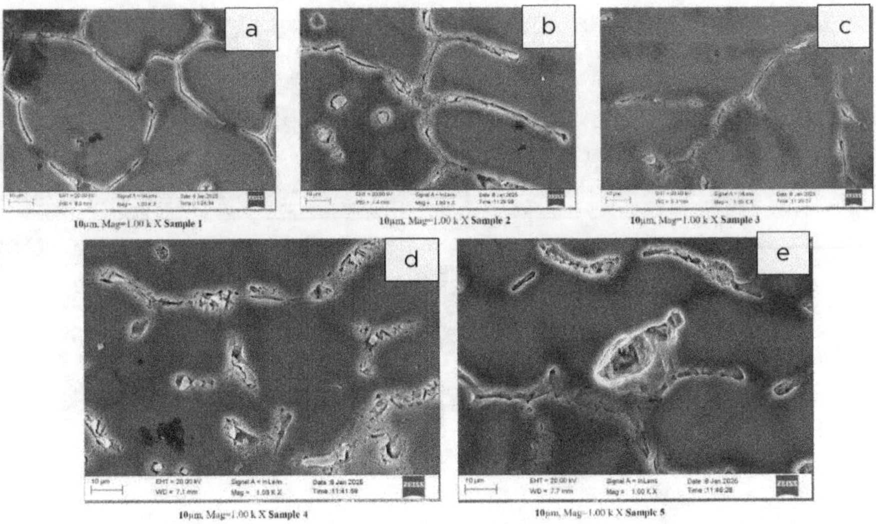

Figure 4.20 FESEM Images for Al 7068, 10 µm, Mag: 1.00 KX (a) Al7068 alloy [S0] (b) Al7068/(1% SiC +1 % Si_3N_4) [S1], (c) Al7068/(0.5% SiC +1.5% Si_3N_4) [S2], (d) Al7068/(0.5% SiC +2.5% Si_3N_4) [S3] & (e) Al7068/(1.0% SiC +2.0% Si_3N_4) [S4]

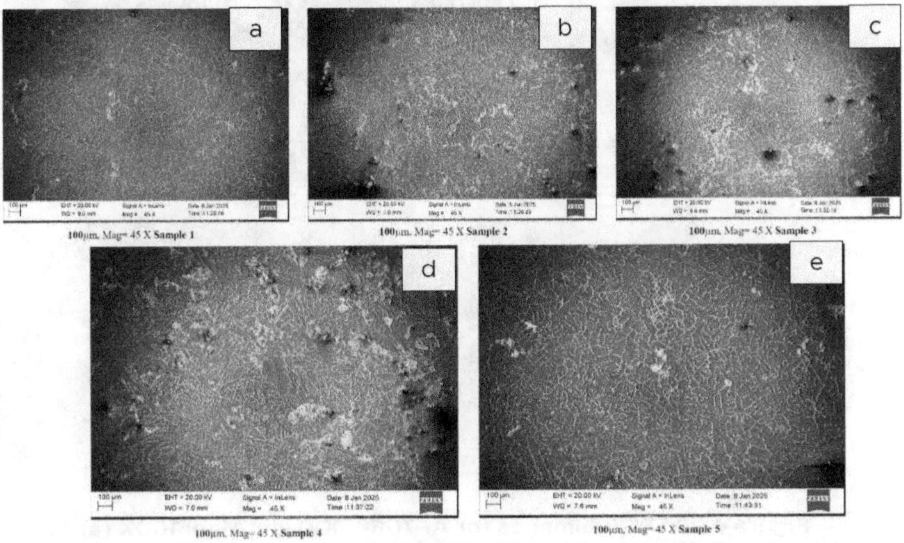

Figure 4.21 FESEM Images for Al 7068, 100 µm, Mag: 45X (a) Al7068 alloy [S0] (b) Al7068/(1% SiC +1 % Si_3N_4) [S1], (c) Al7068/(0.5% SiC +1.5% Si_3N_4) [S2], (d) Al7068/(0.5% SiC +2.5% Si_3N_4) [S3] & (e) Al7068/(1.0% SiC +2.0% Si_3N_4) [S4],

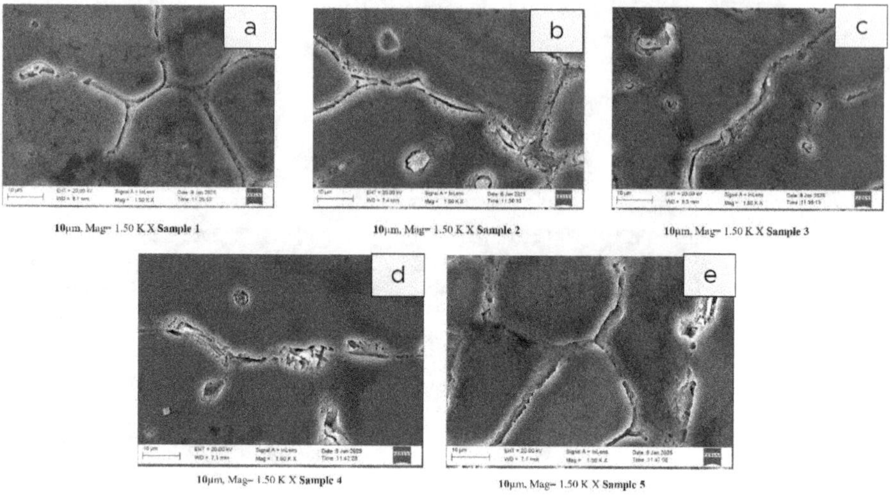

Figure 4.22 FESEM Images for Al 7068, 10 μm, Mag: 1.50 KX (a) Al7068 alloy [S0] (b) Al7068/(1% SiC +1 % Si_3N_4) [S1], (c) Al7068/(0.5% SiC +1.5% Si_3N_4) [S2], (d) Al7068/(0.5% SiC +2.5% Si_3N_4) [S3] & (e) Al7068/(1.0% SiC +2.0% Si_3N_4) [S4],

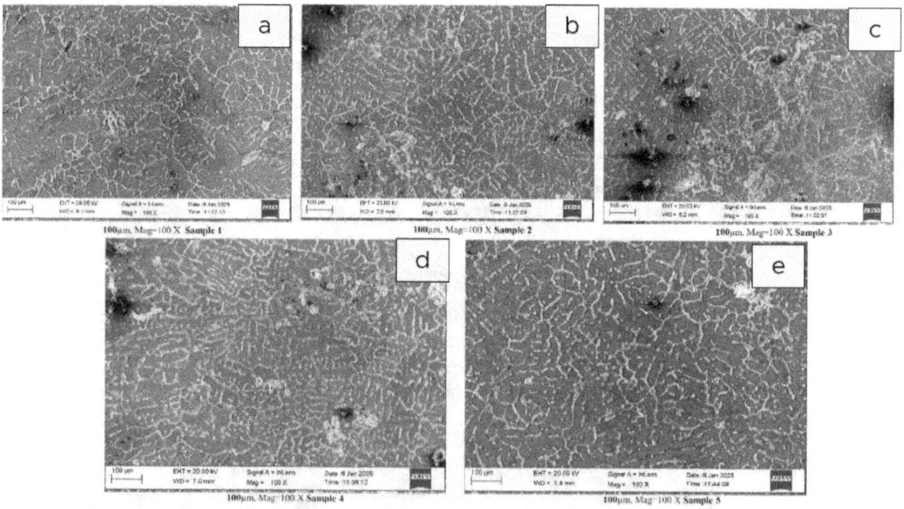

Figure 4.23 FESEM Images for Al 7068, 100 μm, Mag: 100X (a) Al7068 alloy [S0] (b) Al7068/(1% SiC +1 % Si_3N_4) [S1], (c) Al7068/(0.5% SiC +1.5% Si_3N_4) [S2], (d) Al7068/(0.5% SiC +2.5% Si_3N_4) [S3] & (e) Al7068/(1.0% SiC +2.0% Si_3N_4) [S4]

Table 4.4 FESEM Microstructural Analysis of Al7068 and Al7068/SiC/Si_3N_4 Composite Samples

Sample	SEM Magnification	Observed Microstructural Features	Interpretation and Significance
Sample1 (Base Al7068)	45×	Relatively smooth surface with clearly defined grain boundaries and minimal surface irregularities	Indicates a homogeneous aluminum matrix with coarse grains and limited strengthening mechanisms
	300×	Large and continuous grains with absence of secondary phases	Typical microstructure of unreinforced aluminum alloy, resulting in lower mechanical strength
	700×–1000×	Smooth grain boundaries and weak obstruction to dislocation motion	Reflects limited resistance to deformation and crack propagation
Sample 2	45×	Slight increase in surface roughness with finer grain network	Initial grain refinement due to reinforcement addition
	300×–500×	Fine reinforcement particles visible along grain boundaries	Particles act as barriers to dislocation movement, improving strength
	700×–1000×	Good matrix–particle interfacial bonding with minimal defects	Indicates effective load transfer between matrix and reinforcement
	45×	Moderately rough surface with uniformly distributed microstructural features	Suggests improved dispersion of reinforcement particles
Sample 3	300×–500×	Refined grain structure with evenly distributed reinforcement	Grain refinement enhances mechanical performance
	700×–1000×	Strong interfacial continuity and minimal porosity	Promotes improved toughness and crack resistance
Sample 4	45×	Uneven surface morphology with visible reinforcement-rich regions	Indicates non-uniform particle distribution
	300×–500×	Particle agglomeration and micro-voids near grain boundaries	Agglomeration leads to stress concentration sites
	700×–1000×	Presence of micro-cracks and interfacial discontinuities	Suggests reduced structural integrity and lower mechanical reliability
	45×	Uniform surface texture with fine grain network	Demonstrates effective processing and optimal reinforcement distribution
Sample 5	300×–500×	Highly refined grains with homogeneously dispersed reinforcement	Enhances strength through grain boundary strengthening

700×–1000×	Clean interfaces, minimal porosity, and strong matrix–reinforcement bonding	Indicates superior load transfer and resistance to crack initiation

4.2.2 Optical Microscopic Analysis

Microstructural studies were conducted by using Carl- Zeiss Axio Scope 5 at MANIT, Bhopal. Figure 4.24 presents optical micrographs image of the Al7068 alloy and its composites reinforced with varying proportions of silicon nitride (Si_3N_4) and silicon carbide (SiC). The microstructural changes in the samples from S0 to S1 showing the influence of silicon nitride (Si_3N_4) and silicon carbide (SiC) micro particles on grain morphology, distribution uniformity, and phase interaction. This optical micrograph images were captured at 100× magnification, corresponding to 100 μm.

The unreinforced Al7068 alloy exhibits a coarse dendritic α-Al microstructure with prominent grain boundaries and intermetallic networks. The absence of reinforcements or without the mixing of micro particles results in larger, irregular dendrite arms due to unimpeded grain growth during solidification as shown in Figure 4.24 (a). Mixing of silicon nitride (Si_3N_4) and silicon carbide (SiC) micro particles become showing noticeable refinement in dendrite size and morphology. The presence of Si_3N_4 and SiC particles showing heterogeneous nucleation, resulting in finer, more uniformly distributed grains. However, dark agglomerations suggest partial clustering of Si_3N_4 reinforcements as present in Figure 4.24 (b). The Figure 4.24 (c). represents with a higher proportion of Si_3N_4 and small proportion of SiC, the structure exhibits increased grain refinement and matrix disruption. The SiC particles effectively restrict dendritic growth, though micro voids and black patches indicate potential porosity and particle agglomeration during processing. This composition Figure 4.24 (d). showing the most refined and fragmented dendritic network among all samples. SiC, being a potent grain refiner, promotes a dense distribution of fine grains and disrupted dendrite arms. The matrix-reinforcement interface appears more coherent, suggesting improved load-bearing characteristics. Some isolated porosity is visible, likely due to high reinforcement content. The Figure 4.24 (e). represents the increasing Si_3N_4 content further modifies the morphology, producing a semi-equiaxed grain structure with interrupted dendritic arms. The higher total ceramic content results in increased nucleation sites, enhancing uniformity. This analysis confirms that mixing of silicon nitride (Si_3N_4) and silicon carbide (SiC) micro particles have been improved the Al7068 alloy mechanical behavior by tailoring the microstructural features.

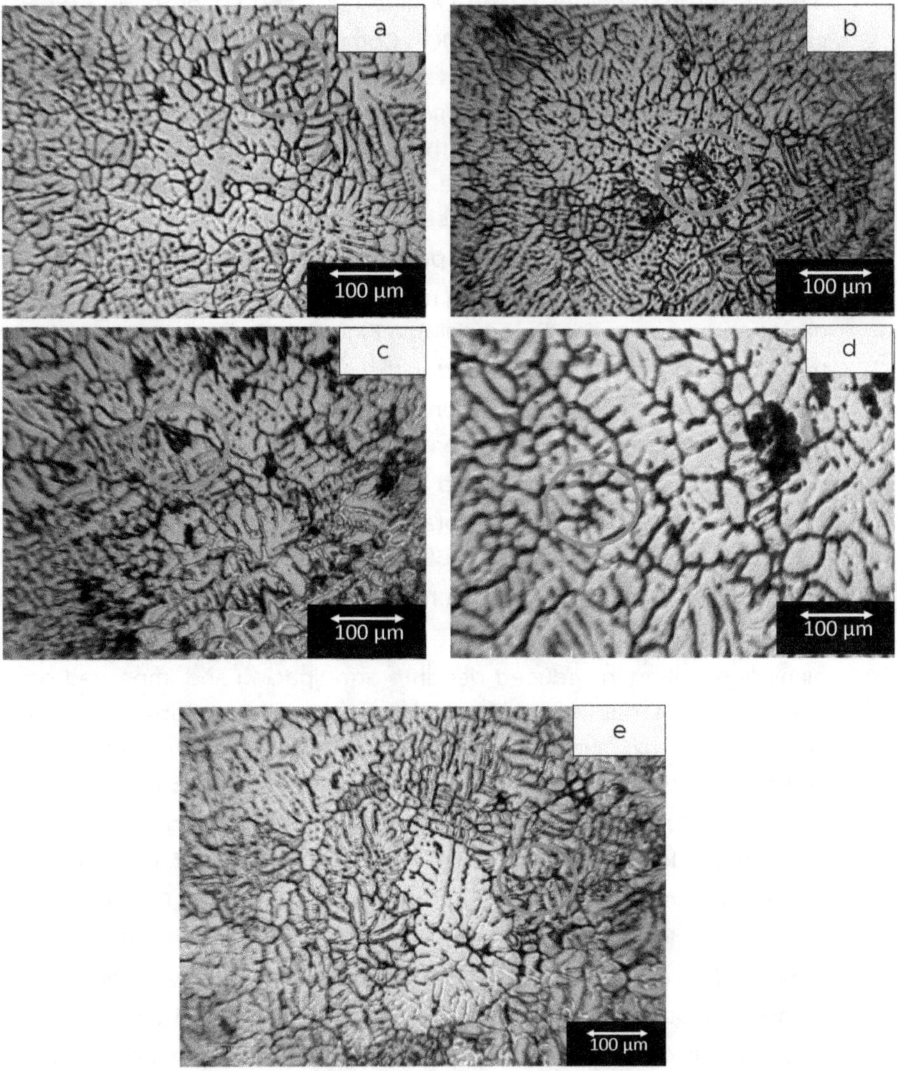

Figure 4.24 Optical Micrographs Images (Mag=100X) for Al7068 (a) Al7068 alloy [S0] (b) Al7068/(1% SiC +1% Si_3N_4) [S1], (c) Al7068/(0.5% SiC +1.5% Si_3N_4) [S2], (d) Al7068/(0.5% SiC +2.5% Si_3N_4) [S3] & (e) Al7068/(1.0% SiC +2.0% Si_3N_4) [S4]

Figure 4.25 showed the optical micrographs images of the unreinforced base alloy Al7068 and its hybrid metal matrix composites reinforced with different proportions of silicon carbide (SiC) and silicon nitride (Si_3N_4), captured at a magnification of 500×. Compared to lower magnifications, these images provide enhanced clarity of grain boundaries, interdendritic regions, and reinforcement–matrix interactions. Figure 4.25(a) represents the microstructure of the unreinforced Al7068 alloy (S0). The structure is dominated by coarse, irregular dendritic grains with clearly visible grain boundaries. The absence of reinforcement silicon carbide (SiC) and silicon nitride (Si_3N_4) particles allows unrestricted dendritic growth during the solidification of metal, resulting in comparatively larger grains and a non-refined morphology. Figure 4.25 (b), corresponding to the Al7068 composite reinforced with 1% SiC and 1% Si_3N_4 (S1), shows noticeable grain refinement when compared with the base alloy. The presence of fine, dark reinforcement silicon carbide (SiC) and silicon nitride (Si_3N_4) particles distributed along grain boundaries and within the matrix indicates effective incorporation of the ceramic phases. In Figure 4.25 (c), which depicts the Al7068 composite containing 0.5% SiC and 1.5% Si_3N_4 (S2), the microstructure appears further refined and more uniform. The increased content of Si_3N_4 enhances grain boundary pinning, resulting in reduced dendrite arm spacing and improved grain uniformity. The reinforcement particles are well embedded within the aluminum matrix, suggesting good interfacial bonding. Figure 25(d) corresponds to the Al7068 composite reinforced with 0.5% SiC and 2.5% Si_3N_4 (S3). This micrographs image shows the fine grain structure, localized dark areas indicated slight agglomeration of reinforcement silicon carbide (SiC) and silicon nitride (Si_3N_4) particles. Figure 4.25 (e) shows the microstructure of the Al7068 composite with 1.0% SiC and 2.0% Si_3N_4 (S4). This sample exhibited the highly refined and homogeneous grain structure with uniformly distributed silicon carbide (SiC) and silicon nitride (Si_3N_4) micro particles. The well-balanced mixing of SiC and Si_3N_4 by stirring operation in sample 5 refined and uniform the metal morphology and to improve the tensile strength, hardness, and impact resistance.

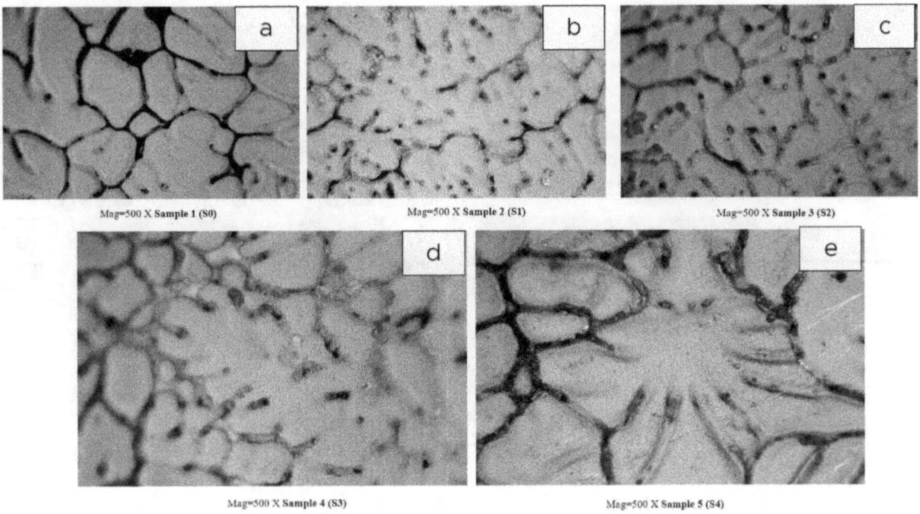

Figure 4.25 Optical Micrographs Images (Mag=500X) for Al7068 (a) Al7068 alloy [S0] (b) Al7068/(1% SiC +1% Si_3N_4) [S1], (c) Al7068/(0.5% SiC +1.5% Si_3N_4) [S2], (d) Al7068/(0.5% SiC +2.5% Si_3N_4) [S3] & (e) Al7068/(1.0% SiC +2.0% Si_3N_4) [S4]

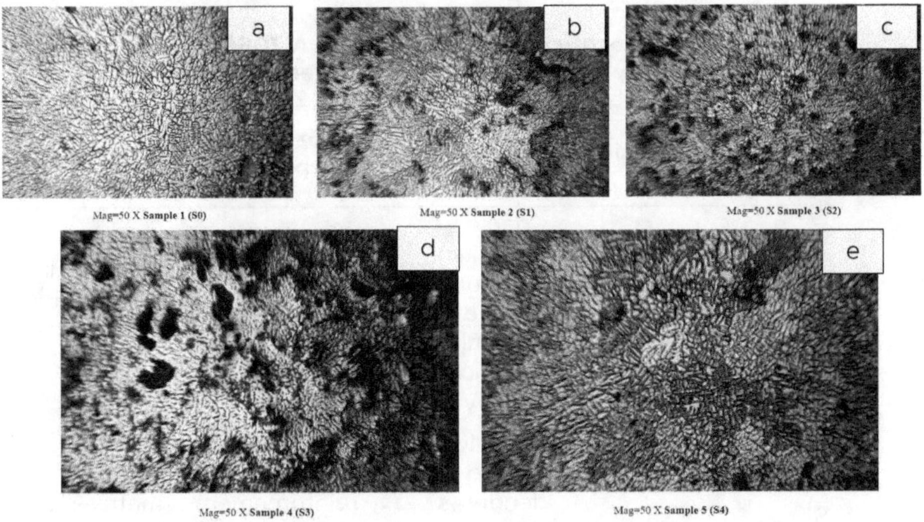

Figure 4.26 Optical Micrographs Images (Mag=50X) for Al7068 (a) Al7068 alloy [S0] (b) Al7068/(1% SiC +1% Si_3N_4) [S1], (c) Al7068/(0.5% SiC +1.5% Si_3N_4) [S2], (d) Al7068/(0.5% SiC +2.5% Si_3N_4) [S3] & (e) Al7068/(1.0% SiC +2.0% Si_3N_4) [S4]

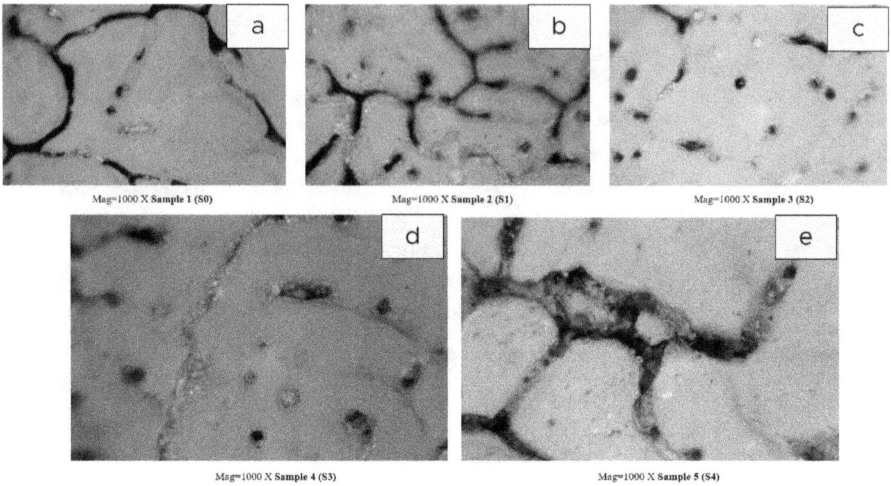

Figure 4.27 Optical Micrographs Images (Mag=1000X) for Al7068 (a) Al7068 alloy [S0] (b) Al7068/(1% SiC +1% Si_3N_4) [S1], (c) Al7068/(0.5% SiC +1.5% Si_3N_4) [S2], (d) Al7068/(0.5% SiC +2.5% Si_3N_4) [S3] & (e) Al7068/(1.0% SiC +2.0% Si_3N_4) [S4]

Table 4.5 Microstructural Comparison of Al7068 and Al7068/SiC–Si_3N_4 Hybrid Composites at Different Magnifications

Sample	Composition	50× (Macro-scale features)	500× (Grain morphology)	1000× (Interfacial characteristics)
S0	Al7068	Coarse dendritic structure formed due to non-uniform solidification of alloy	Large equiaxed grains; distinct grain boundaries	Grain boundary segregation and micro-voids observed
S1	Al7068 (1% SiC + 1% Si_3N_4)	Partially refined dendrites happened with improved uniformity	Reduced grain size; reinforcements distributed along alloy boundaries	Improved matrix–particle bonding with limited porosity
S2	Al7068 (0.5% SiC + 1.5% Si_3N_4)	More compact and uniform dendritic structure	Fine equiaxed grains; homogeneous distribution of	Strong interfacial bonding and effective grain

		formed as compared to S1	reinforcement	boundary pinning
S3	Al7068 (0.5% SiC + 2.5% Si_3N_4)	Disrupted dendritic pattern is found with localized clustering	Significant grain refinement with agglomeration zone	Particle clustering and micro-porosity evident
S4	Al7068 (1.0% SiC + 2.0% Si_3N_4)	Dense and uniform microstructure formed with minimal dendrites	Highly refined grains found with continuous boundaries	Excellent matrix–reinforcement adhesion and negligible defects

4.2.3 Fractographic Analysis of Tensile Fractured Surface

The SEM fractographic Figure 4.26 (a–d) provides insight into the material's failure mechanisms and correlates with the mechanical properties observed in earlier graphs.

Figure 4.28(a) shows prominent microvoids, fibrous tendons, and microcracks—features indicative of ductile fracture. Inspection of the fracture surface reveals the presence of large, shallow microvoids distributed throughout the surface. The surface morphology, appearing rough and fibrous, reflects a typical cup-and-cone ductile fracture. These characteristics suggest significant plastic deformation occurred before failure, consistent with the higher strength and lower elongation observed in sample S1 as compared to S0. Microvoid coalescence and fibrous morphology indicate a classic ductile failure mode, where stress is distributed across the matrix before crack propagation occurs.

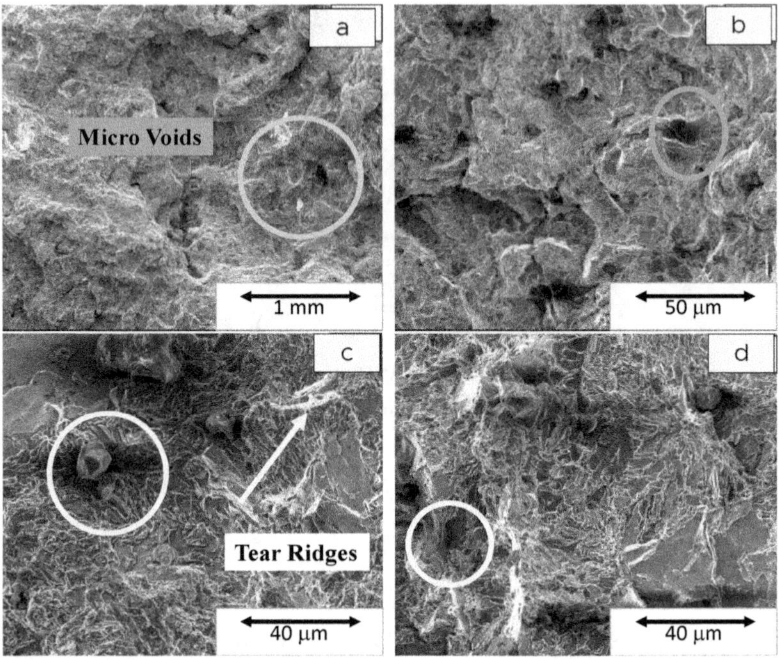

Figure 4.28 FESEM image of tensile test fractured surface for (a) Al7068 alloy (1% SiC +1% Si_3N_4) [S1] (b) Al7068 (0.5% SiC +1.5% Si_3N_4) [S2], (c) Al7068 (0.5% SiC +2.5% Si_3N_4) [S3], and (d) Al7068 (1% SiC +2% Si_3N_4) [S4] samples, respectively

Figure 4.28 (b) reveals tear ridges and evidence of particle fracture and microvoids. Tear ridges suggest higher resistance to crack propagation, while particle fractures indicate stronger matrix-particle interactions. Compared to S1, the fracture surface shows shallower and smaller microvoids, indicating a reduction in ductile deformation due to the presence of hard ceramic particles. The presence of reinforcements (SiC and Si_3N_4) improves the load-carrying capacity of the matrix. Due to weak interfacial bonding at some locations, the circled regions show small voids or cavities formed. This microstructure likely corresponds to S2, where the ultimate tensile strength (730 MPA) peak but restricts elongation. It reflects a more balanced fracture mechanism—neither too brittle nor overly ductile—supporting strength and reasonable ductility.

Figure 4.28 (c) The formation of dense tear ridges shows that small plastic deformation occurred in localized regions. These ridges show that energy was absorbed before the propagation of the crack, highlighting the mixed-mode fracture. Compared to S1, the circled dimples appear smaller and more uniform, indicating improved crack resistance and a refined microstructure from proper reinforcement dispersion. when the reinforcement content increases, some regions indicate little fracture of Si_3N_4 particles. The presence of finer dimples and tear ridges indicates a partially brittle fracture mode. As the SiC content increases, it enhances the stiffness and hardness. The fracture mechanism happens in such a way that shows both toughness and brittleness by strong interfacial bonding, improved load transfer, and fewer micro voids grow. This S4 surface morphology reflects improved tensile strength (752 MPA), yield strength (374 MPA) and hardness (65.57 HRB) as compared to S3, S2 and S1.

Chapter - 5
Conclusion and Future Scope

Chapter 5
Conclusion and Future Scope

5.1 Conclusion

In accordance with the stated research objectives, the following aims have been successfully achieved:

a) Al7068 alloy reinforced with SiC and Si_3N_4 microparticles have been successfully developed by using the stir casting technique.

b) The addition of dual reinforcements resulted a significant enhancement in mechanical properties, including tensile strength, yield strength, hardness, and toughness.

c) After T6 heat treatment process further improved the mechanical properties but decreased the ductility.

d) Microstructural analyses (Optical Microscopy & FESEM) confirmed that uniform dispersion of SiC and Si_3N_4 microparticles and enhanced resistance to crack initiation and propagation.

e) Fractographic analysis indicated that transition from predominantly ductile to mixed-mode fracture behavior, confirming improved load transfer and interfacial bonding in the developed composites.

The study comprehensive findings and ensuing conclusions can be summarized as follows:

1) The variation in UTS across different samples demonstrates the positive impact of hybrid reinforcement. Sample S4 achieved the highest optimum tensile strength of 752 MPa, marking a remarkable 18.43% improvement over the base alloy (SO). This significant improvement is achieved by the combination of reinforcing effect of silicon carbide (SiC) and silicon nitride (Si_3N_4) particles, which, when integrated with optimized processing techniques, led to enhanced strength characteristics.

2) There is some variation in ultimate tensile strength (UTS) in both the as cast and heat-treated condition for samples SO, S3, and S4. It's observed that after heat treatment, all the Al7068 compositions show clear improvement in that strength, demonstrating the beneficial role of thermal processing in strengthening Al7068 and its hybrid composites.

3) Ultimate tensile strength (UTS) increases from 635 MPa in the as-cast condition to 656 MPa after heat treatment process for the base alloy (S0), reflecting improved matrix continuity and stress redistribution. Sample S3 exhibits the highest UTS from all the tested specimen, rising from 734 MPa to 736 MPa, which indicates an optimal value of reinforcement combination that effectively restricts dislocation motion. Similarly, sample S4 (1.0% SiC + 2.0% Si_3N_4) shows a noticeable improvement from 752 MPa to 761 MPa following heat treatment.

4) The ultimate tensile strength results clearly demonstrate that heat treatment has positive effect on the mechanical behavior of Al7068 alloy and its hybrid composites. All investigated samples showing higher ultimate tensile strength (UTS) values under the heat-treated condition. From all tested specimen, sample S4 gained the maximum tensile strength properties, highlighting the effectiveness of well-balanced reinforcement content in strengthening the Al7068 alloy. Heat treatment reduces casting-related defects and internal residual stresses while promoting a more uniform dispersion of reinforcements, thereby improving load transfer efficiency during tensile deformation.

5) The variation in yield strength (YST) of samples S0 to S4. A progressive increase in YST is observed from the base alloy S0 to sample S4. The yield strength of the unreinforced Al7068 alloy (S0) is recorded as 310 MPa, which increases steadily to 324 MPa in S1 and 365 MPa in S3, reaching a maximum value of 374 MPa in S4. The presence of hard reinforcements contributes to load-bearing capability and grain refinement, resulting in improved yielding behavior. The superior performance of S4 showing an optimal reinforcement composition that maximizes strengthening mechanisms without introducing adverse microstructural effects.

6) The yield strength study clearly shows that silicon carbide (SiC) and silicon nitride (Si_3N_4) microparticles content has done significant improvement on the yielding behavior of Al7068-based composites. Yield strength improves by 20.65 % progressively from S0 to S4, with sample S4 (1% SiC +2% Si3N4) exhibiting the highest resistance to plastic deformation. This research outcomes confirms that an optimized reinforcement combination effectively enhances Al7068 alloy strengthening and load transfer efficiency.

7) Overall, from all the sample it concludes that, sample S4 (1% SiC +2% Si_3N_4) validates the most satisfactory yield strength properties, making it the most suitable composition for the applications of high resistance to plastic deformation and reliable structural mechanical performance.

8) The variation of percentage elongation reveals that mixing of SiC and Si_3N_4 microparticles decreases the elongation from sample S0 to S4, with sample S4 (1% SiC +2% Si_3N_4) achieving the lowest elongation of 9.35% showing decrement over the base alloy (S0 at 15.2%). As the elongation decreases it ultimately increases the material hardness and strength. So, the results demonstrate clear inverse relationship between strength and ductility.

9) The comparison of "as-cast" and T6 heat-treated samples (S0, S3, and S4) clearly shows that T6 heat treatment reduces percentage elongation from 15.2% to 13.8% (S0), 11.6% to 10.2% (S3), and 9.35% to 9.1% (S4), indicating that improved ultimate tensile strength, yield strength but ductility has been reduced. The observed mechanical behavior indicates microstructural refinement, improved homogeneity as well as relief of internal stresses during the heat treatment process. Therefore, T6 heat treatment enhances the material strength.

10) The hardness of Al7068 composites improves consistently with the increased the reinforcement by 1% SiC & 2 %Si_3N_4. The combination of Si_3N_4 and SiC enhances the mechanical strength of the matrix through grain refinement and improved matrix-particle interaction. Sample S4 show the best-balanced composition of SiC & Si_3N_4. showing increased hardness of 54.61% (65.57 HRB) as compared to base alloy Al7068 (S0). Therefore, the greater hardness is observed in sample S4 also shows enhanced tensile strength and mechanical stability as well as it's highly suitable for applications requiring high wear resistance, surface durability, and improved structural performance.

11) Post heat treatment process on all the samples of Al7068 exhibited notable increase in the properties of hardness, with Sample S4 achieving the maximum value of 104.6 HRB. Compared to the base alloy's as-cast hardness, the hardness increased by 54.64%. Sample

S4 is showing most effective reinforcement (1.0% SiC + 2.0% Si_3N_4) composition to improving the surface hardness properties as well as making it highly suitable for applications requiring mechanical strength.

12) Reinforcement silicon carbide (SiC) and silicon nitride (Si_3N_4) is very important microparticles to improving the mechanical properties of Al7068 composites alloy, specially under the action of impact loading. Sample S4 exhibited the highest impact toughness, absorbing 2.0 J of energy, which corresponds to a 100% increase over the base alloy. highly suitable for structural and aerospace applications where excellent toughness is required.

13) After shot peening operation on the Charpy specimen, it improves the impact toughness of Al7068 alloy. Sample S4 showed the maximum improvement increase by 116.67% absorbed energy compared to base alloy S0, indicating excellent resistance under dynamic loading. These outcomes results proposed that microstructural changes happened by shot peening process, with reinforcement effects, effectively inhibit crack propagation and enhance energy absorption. Therefore, shot peening process proves that it is a valuable post-processing technique to strengthen the aluminum matrix composites for structural and impact-critical applications.

14) SEM results show that adding reinforced particles improves the Al7068 microstructure by making the grains finer and improved particle dispersion. When the reinforcement amount is added by 0.5% SiC & 2.5% Si_3N_4 the structure becomes uniform and stronger. But when balanced reinforcement is added by 1% SiC & 2 % Si_3N_4 particles start to cluster together. This clustering may increase hardness, toughness but it can reduce ductility.

15) The unreinforced alloy exhibits coarse dendritic structure with less grain refinement, whereas the more mixing of silicon carbide and silicon nitride progressively modifies the Al7068 morphology. At lower reinforcement content, grain refinement starts through heterogeneous nucleation, leading to a disruption of the dendritic network. With additional mixing of silicon carbide and silicon nitride, the microstructure becomes more refined, and particle dispersion improved. The optimum reinforcement condition demonstrates a

uniform and homogeneous distribution of Si_3N_4 and SiC particles with a largely equiaxed grain structure, indicating effective matrix–reinforcement bonding and enhanced load transfer capability. This refined and well-dispersed microstructure is responsible for the observed improvement in mechanical properties. However, excessive reinforcement results in particle clustering and non-uniform dispersion, which may act as stress concentration sites. Due to more clustering of microparticles can increase hardness, it may adversely affect ductility and fracture toughness.

16) Overall, the SEM results confirm that controlled hybrid reinforcement of silicon carbide and silicon nitride microparticles is crucial for achieving a balanced microstructure of Al7068 alloy, to enhanced tensile strength and hardness without impacting structural integrity.

17) The optical microstructural study clearly elaborates that the addition of Si_3N_4 and SiC micro-reinforcements have significant impact on the grain structure and overall morphology of the Al7068 alloy. The unreinforced alloy exhibits a coarse dendritic α-Al structure, whereas the mixing of silicon carbide and silicon nitride progressively refines the dendrites and disrupts the continuous grain network. The presence of Si_3N_4 and SiC generates heterogeneous nucleation during solidification of metal Al7068 alloy, leading to finer and more uniformly distributed grains. Compositions with balanced proportions of SiC and Si_3N_4 show enhanced grain refinement and improved Al7068 matrix–reinforcement bonding, indicating effective load transfer and improved structural integrity. The sample of Al7068 reinforced with silicon carbide and silicon nitride content exhibits the most refined and fragmented dendritic network, which is beneficial for improving mechanical properties of alloy. However, at higher reinforcement content, localized porosity and particle agglomeration are observed in the optical micrograph image, which may adversely affect ductility despite improved grain refinement.

18) Overall, the optical micrographs confirm that controlled hybrid reinforcement with Si_3N_4 and SiC microparticles effectively tailors the microstructure of Al7068 alloy. An optimized combination of reinforcements leads to refined grain morphology and uniform particle distribution, and then contributing to the enhanced mechanical behavior of the composite material.

19) The fracture surface analysis indicates that S1, S2, and S3 samples predominantly ductile failure mode, characterized by tearing ridges, shallow dimples, and fibrous structures. The Al7068 alloy (S1) exhibited typical ductile features with dendritic grains and uniformly distributed dimples, while the composites (S2 and S3) showed refined, shallow dimples, indicating enhanced matrix-reinforcement interaction. However, minor microcracks observed in the S4 sample exhibits superior mechanical performance, achieving a tensile strength of 752 MPa, yield strength of 374 MPa, and hardness of 59.7 HRB, due to its refined microstructure and strong interfacial bonding.

20) Sample-S4 (1% SiC +2% Si_3N_4) is more suitable for aerospace applications because it shows the maximum Tensile Strength. In aerospace structures, materials must carry high static loads, maintain dimensional stability, resist deformation during service & provide high strength-to-weight ratio. Aerospace components such as structural brackets, wing fittings, aircraft interior frames, landing gear accessories, and lightweight support structures require materials with high strength, low density, and excellent fatigue resistance.

21) Also Sample-S4 (1% SiC +2% Si_3N_4) is more appropriate for applications subjected to dynamic and shock loading such as protective structures, automotive components, and impact-resistant panels. In the automotive sector, the composite may be used in braking system components, connecting rods, suspension parts, transmission housings and pistons. The overall results clearly indicate a strength–toughness balance in the developed composites, where Sample-S4 is optimized for high static strength and impact resistance.

5.2 Future Scope

- Future studies can be focused on industrial validation through collaborative research with leading organization. This will help assess its practical performance and promote its use in industrial applications.
- Future research can focus on evaluating the fatigue life and crack growth behavior of Al7068–SiC/Si_3N_4 hybrid composites under cyclic loading conditions.

- Corrosion resistance in aggressive environments and high-temperature mechanical behavior, including creep and thermal stability, should also be investigated to assess their suitability for aerospace and automotive applications.

5.3 Social Impact

5.3.1 Advancement in Material Science:

The research on Al7068 reinforced with silicon carbide (SiC) and silicon nitride (Si_3N_4) significantly contributes to the advancement of material science by analyzing the influence of dual ceramic reinforcements on the mechanical and structural behavior of aluminum matrix composites. The integration of SiC and Si_3N_4 enhances the alloy's strength, wear resistance, and thermal stability, paving the way for the development of next-generation high-performance materials suitable for aerospace, defense, and automotive applications.

5.3.2 Enhanced Mechanical Properties:

The incorporation of SiC and Si_3N_4 reinforcements into Al7068 has demonstrated remarkable improvements in mechanical properties, including peak stress, peak load, hardness, and tensile strength. Silicon carbide contributes to superior wear and load-bearing capacity, while silicon nitride improves fracture toughness and fatigue resistance. These enhancements collectively lead to composites with higher reliability and longer service life, enabling safer and more efficient structural components across engineering sectors.

5.3.3 Potential for Light Weight:

Al7068 composites reinforced with SiC and Si_3N_4 offer an excellent balance of high strength-to-weight ratio and enhanced durability. The lightweight nature of the aluminum matrix, combined with the superior mechanical attributes of ceramic reinforcements, makes these composites ideal for applications in aerospace, automotive, and transportation industries where weight reduction is critical. The resulting decrease in fuel consumption, emissions, and energy usage supports sustainability initiatives and promotes environmentally conscious engineering solution.

www.ingramcontent.com/pod-product-compliance
Lightning Source LLC
LaVergne TN
LVHW010732061126
851857LV00014B/1444
* 9 7 8 1 9 6 2 1 1 6 5 4 1 *